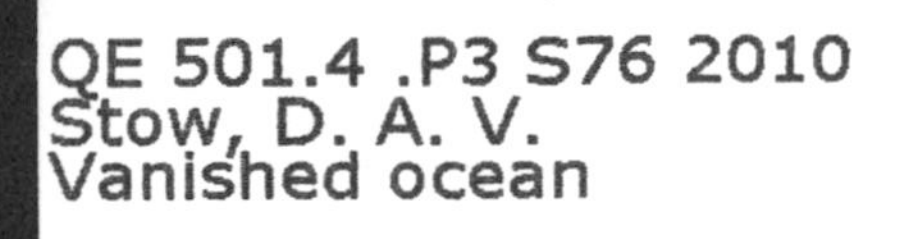
QE 501.4 .P3 S76 2010
Stow, D. A. V.
Vanished ocean
W9-BSG-424

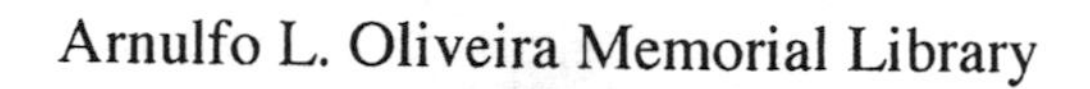
Arnulfo L. Oliveira Memorial Library

WITHDRAWN
UNIVERSITY LIBRARY
THE UNIVERSITY OF TEXAS RIO GRANDE VALLEY

LIBRARY
THE UNIVERSITY OF TEXAS
AT BROWNSVILLE
Brownsville, TX 78520-4991

VANISHED OCEAN

Vanished Ocean

HOW TETHYS RESHAPED THE WORLD

DORRIK STOW

OXFORD
UNIVERSITY PRESS

LIBRARY
THE UNIVERSITY OF TEXAS
AT BROWNSVILLE
Brownsville, TX 78520-4991

Great Clarendon Street, Oxford OX2 6DP

Oxford University Press is a department of the University of Oxford.
It furthers the University's objective of excellence in research, scholarship,
and education by publishing worldwide in

Oxford New York

Auckland Cape Town Dar es Salaam Hong Kong Karachi
Kuala Lumpur Madrid Melbourne Mexico City Nairobi
New Delhi Shanghai Taipei Toronto

With offices in

Argentina Austria Brazil Chile Czech Republic France Greece
Guatemala Hungary Italy Japan Poland Portugal Singapore
South Korea Switzerland Thailand Turkey Ukraine Vietnam

Oxford is a registered trade mark of Oxford University Press
in the UK and in certain other countries

Published in the United States
by Oxford University Press Inc., New York

© Dorrik Stow 2010

The moral rights of the author have been asserted
Database right Oxford University Press (maker)

First published 2010

All rights reserved. No part of this publication may be reproduced,
stored in a retrieval system, or transmitted, in any form or by any means,
without the prior permission in writing of Oxford University Press,
or as expressly permitted by law, or under terms agreed with the appropriate
reprographics rights organization. Enquiries concerning reproduction
outside the scope of the above should be sent to the Rights Department,
Oxford University Press, at the address above

You must not circulate this book in any other binding or cover
and you must impose the same condition on any acquirer

British Library Cataloguing in Publication Data
Data available

Library of Congress Cataloging in Publication Data
Library of Congress Control Number: 2010922420

Typeset by SPI Publisher Services, Pondicherry, India
Printed in Great Britain
on acid-free paper by
Clays Ltd., St Ives plc

ISBN 978–0–19–921428–0

1 3 5 7 9 10 8 6 4 2

FOR

Claire

3/7/11 UTB Baker & Taylor $23.96

CONTENTS

PREFACE

This is the story of a lost ocean, which is known to geologists as the Tethys Ocean. It lasted for 250 million years of Earth history, dominating the equatorial world and playing host to the changing life and events that have shaped the world we inhabit today. As continents moved and sea levels rose, Tethys waters swept north across large tracts of Europe, Asia and North America, and south over Africa and South America. As sea level fell so Tethys receded. Continent ground against continent and Tethys was finally squeezed out of existence just 5½ million years ago. But there is a very rich legacy from this past ocean hidden in the rocks of many continents and buried deep beneath the ocean floors of the present-day world.

For many years of my professional career as a geologist and oceanographer, I have worked on rocks on land or drilled into sediments beneath the seas that were once a part of the Tethys Ocean. Slowly and carefully, I have gathered countless clues and amassed a large body of evidence. Sometimes I was looking directly for such evidence, but equally often I was working on some altogether different topic when a new piece of the jigsaw puzzle appeared and neatly slotted into place. That is the way of science. A great many other scientists over many years have been involved, directly or indirectly, in research on the Tethys. The results of their research, like my own, are published in scientific journals and books across the world.

The evidence is incontrovertible. This is a true story – or at least as true a rendition of the scientific facts as we can produce from our current understanding. It is not some fanciful musing or creative science fiction, but as close to an accurate account of Tethys history as I can make it. And yet there were volcanic eruptions without parallel in today's world; mass extinctions that almost ended all life on Earth; dramatic radiation of new species in a fecund ocean known only from their tantalizing fossil remains; 'black death' in the Tethys that led to the world's richest oil reserves; and a time when oceans rose up and spread across the planet until only 18% remained as dry land.

All this and more, I have tried to capture as a storyline through time. The book starts with an explanation of the sorts of clues we can find in the rocks, and the information they can provide, and introduces the broad workings of plate tectonics and the all-important measurement of time. Each subsequent chapter then presents a new period of the Tethys, from the building of the supercontinent of Pangaea that marks the beginning of Tethys to its eventual demise as a gaping hole and blistering salt inferno somewhere between Africa and Europe. I have used maps judiciously through the text to help paint a portrait of evolution and change. In particular, the maps at the beginning of each chapter show a reconstruction of the world and the Tethys Ocean for one snapshot of the time covered in that chapter. There is a geological timescale and Tethys timeline in Chapter 1 to help the reader master the huge timescales involved and the names that geologists use as shorthand for different periods in the past. There is also a brief glossary of the more obscure technical terms, which I have introduced of necessity through the text.

In addition to my own research, the sources of scientific information used in reconstructing this history are altogether too numerous to detail in a book of this nature. I have therefore decided to abandon the idea of having an endless series of footnotes or endnotes referring to specific scientific articles. Instead, there is a short bibliography with reading suggestions for those wishing to delve further into the science behind the story. I have also introduced in the text just a few of the pioneers in geological science, whose work has made this historical reconstruction possible, as well as some of the many colleagues and friends who have helped me in the field and at sea.

August 2009 Dorrik Stow

ACKNOWLEDGEMENTS

Claire – thank you for your love and support always, for your understanding and companionship, for your constant joy and enthusiasm in the field, and for being my wife. Thank you also for your insightful comments on an earlier version of the manuscript. To my children – Jay, Lani and Kiah – thank you for your love and youthful wisdom, for tolerating the many field trips and also for critically reading the manuscript. To my parents – Terry and Jill – thank you for everything.

I gratefully acknowledge the various institutes in which I have worked that have wittingly or unwittingly lent their support to my Tethyan research. These include: the universities of Edinburgh, Nottingham and Southampton in the UK; Dalhousie University in Canada, Université de Bordeaux in France; The British National Oil Corporation in Glasgow; British Petroleum in London; and the Instituto de Oceanografia de Espanol in Malaga. Lastly and, most importantly, my full appreciation to Heriot Watt University in Edinburgh, where I currently work as the Ecosse Director and which has provided full technical and secretarial support as well as intellectual stimulation. John Wright expertly drafted the paleoworld maps that open each chapter.

I have made many visits to other institutes around the world and gained huge support from a great number of colleagues. I have been very fortunate in having research students and assistants who have taught me at least as much as I have taught them. Only some of these colleagues and past students are mentioned in text, but my warmest thanks go to all. Mounting research investigations on land and especially at sea often involves complex and costly logistical support for which I am greatly indebted. Financial support has been from a wide range of national, international and independent sources. All of these have been formally acknowledged before in specific research publications, but I should like to record my collective thanks for their part in enabling this work to come before a wider audience.

Of course, any work such as this is a true composite of research and ideas from many different sources, in addition to my own. I fully acknowledge this and take final responsibility for the interpretations I have made.

Finally, my warm thanks and appreciation to the full team at Oxford University Press, their illustrators and their freelance copy-editor, Paul Beverley; and to Latha Menon and Emma Marchant, in particular, for their support, patience and encouragement.

1

Tethys the Sea Goddess

from what mother did they come,
from what volcanic sperm,
oceanic, overflowing,
from what flower before, from which scent
snuffed short by the glacial stare?

From *Stones of the Sky* by Pablo Neruda
(translated by James Nolan)

Part of a Roman floor mosaic depicting Tethys and her husband Okeanos, with the River Dragon between them. Turkey, circa 1st–2nd century CE. (© Murat Sen/iStockphoto.com)

The ancient Greeks knew well the supreme significance of the sea in their time – for trade and conquest, for fishing and philosophy – although none would have truly believed the remarkable tale of a *vanished ocean* to which their sea goddess, Tethys, would unwittingly lend her name some 2000 years later. The mysterious and enchanting story of Tethys, the most powerful sea goddess of Greek legend, is rich with the signature hallmarks that so characterize many colourful tales of that past era of resplendent mythology: tragedy, intrigue, shock and romance. So too the tale that follows. Scientists of today, geologists and oceanographers, have spent many years looking far back into the Earth's history to prove the former existence of what they name the Tethys Ocean. This is an amazing detective story wrought from a diverse series of clues locked away in rocks now exposed high up in mountain ranges and buried in sediments deep beneath our present seas.

The Tethys Ocean once dominated the Earth. Its vast waters bore witness to many of most dramatic episodes in the story of our planet. It played host to a plethora of fascinating organisms, saw great mass extinctions and ultimately nurtured the ensuing rebirth of new life forms. It endured its own secret drama of submarine quakes and powerful currents, at one point quaking under evening skies lit by the colossal outpourings of a supervolcano exploding along its southern margin. For long ages, its quiet central gyres were more desolate and empty than anything seen in today's world. But then, 5½ million years ago, Tethys disappeared.

Many people have heard of the 'Lost Continent of Atlantis', although fewer can say whether it is purely derived from the annals of myth and legend, handed down from generation to generation, or whether there is any scientific basis for the concept. But a 'vanished ocean'? Surely that is far more unlikely! How could an ocean simply vanish? And how could we possibly know it was ever there in the first place? These are the sorts of comments I have heard from my erudite friends outside the sciences in response to my musings about the lost Tethys. To both questions, the remarkable and enchanting answer is that oceans are indeed born from the midst of dusty continents, they grow into wide and wonderful seas, and are then slowly consumed once more beneath an overriding landmass. They leave behind all manner of clues to their former presence and of their long-gone inhabitants in the rock record.

In fact, this record unravelled through years of painstaking geological and oceanographic research reveals that numerous past oceans have come and gone during that inconceivable expanse of time since planet Earth was born some 4½ billion years ago – the Panthalassa, Rheic and Iapetus Oceans, to name but a few. Tethys was the last great ocean on Earth, before those of today's world, and it silently slipped from view at about the same time as our primitive ancestors had begun to walk upright along its shoreline. In Greek mythology, Tethys was married to Okeanos, a mighty river that stretched around the world – a fitting name, which captures something of the greatness, enchantment and elusiveness of this lost world.

This book attempts to recreate that lost world, piecing together the many clues now scattered through the rocks that stretch from Morocco to China, from the depths of the Caspian Sea to the highest Himalayan peaks. Though imperfect and incomplete, the record is varied and captivating, telling something of the coming together and splintering of continents, the rise and fall of mountain ranges, the changing ocean currents with their link to Earth's climate, the remarkable story of the formation of oil in the Middle East, and much more besides. Such changes in the land and ocean have been strongly instrumental in shaping the colourful pageant of life throughout Earth history, as well as through the past 260 million years during which Tethys has existed.

My intention is to lead the reader along a fairly loose timeline through the history of Tethys, from its inception some 260 million years ago to its demise just 5½ million years ago, and to demonstrate how geologists are able to reconstruct scientifically the ancient past from clues left in the rocks. It is an amazing and intriguing story of a lost, but watery world, and it has never before been told in this way. It will also become apparent that the topics uncovered through this unveiling of Tethys history are far broader in their implication and significance – touching on the great scientific paradigms of plate tectonics and the cyclical nature of Earth patterns, fundamental questions of mass extinction and evolution, and critical issues for the 21st century: energy resources, the environment and climate change. The final chapter attempts to draw lessons from the past that will help us today.

GLOBAL JIGSAW PUZZLE

For a number of years now I have been on the trail of the Tethys Ocean, searching beyond the fanciful musings of mythology for hard scientific evidence. That evidence, though still patchy, can be systematically collated and scientifically reconstructed. It is rather like attempting some global jigsaw puzzle from which many of the pieces are missing. Some of the pieces that do exist I have gathered from my own research, while many more rely on the work of other scientists and colleagues, too numerous to mention, from around the world.

The clues are now all jumbled together, as I have found here in Andalucia, that gloriously hot and richly traditional part of southern Spain, where my search is nearly finished. As I begin to order the data, trying to recreate that loose timeline through Tethys history and so begin my unusual tale, I am working in the *Instituto Espanol de Oceanografia,* enjoying one of those wonderful latter-day university traditions we call an extended sabbatical leave from my own department. The institute lies at the end of the harbour pier with its toe firmly dipped in the tranquil Mediterranean Sea. From this vantage point I can gaze out beyond the timeless scenes of nets laid out to dry and awaiting repair, the stacked lobster traps and gently swaying fishing boats, to the wrinkled surface of azure blue. Deep below that surface there is a hidden suture line of immense geological dimension and significance, an annealed scar in the ocean crust, beneath which the Tethys Ocean disappeared for ever.

Behind me I can sense the chaos and furore of a burning August day on Fuengirola beach – cold beer and colourful umbrellas, barbecued seafood from countless *pescadoritas,* children, sandcastles and scantily clad, well-oiled bodies. Still further inland lie the half-worn yet still jagged peaks of the Sierra Nevada, the crumpled remains of that last great collision between the African and European continents – fragments of the ocean floor that once lay between them and that were pushed up into mountains rather than being sucked below the suture. And the holidaymakers on the beach... what chance that they have heard of a lost ocean and might believe my incredible tale? I will set the scene, therefore, by introducing some of the clues that we search for and just what sort of evidence they yield. A fuller explanation of the science and of the unfolding story will follow in subsequent chapters.

There are clues in the tectonic organization of the world and in the distribution of earthquakes and volcanic eruptions. A narrow belt of earthquake activity stretches from the Mid-Atlantic Ridge through the Straits of Gibraltar and across the Mediterranean Sea, first confined to the North African margin and then, farther east, as a broader swathe including active volcanoes, and affecting both Europe and Africa. This belt marks the suture line I referred to above, the boundary between two of the vast tectonic plates that make up the Earth's outer layer, or lithosphere, the zone along which they bump and grind against one another. It is the movement of these plates, imperceptibly slow but inexorably forceful, that causes oceans to open up and flourish for a while, but later to narrow and eventually be squeezed out of existence as the direction of plate movement changes. It is the immense friction and heat caused by their relative motion that initiates earthquakes and contributes to the eruption of volcanoes.

A continuation of the same earthquake-volcanic belt passes through Turkey and the Middle East, through central Asia and the Himalayas, everywhere marking the east–west suture line along which the Tethys Ocean eventually disappeared. The terrible earthquakes and violent eruptions along this zone that have stirred, shaken and destroyed people and homes through time are the result of tectonic plates still pushing and grinding against each other. We can chart their occurrence – from the eruption of Santorini that destroyed the Minoan civilization on Crete some 3500 years ago or that of Vesuvius that so cruelly incinerated the thriving Roman town of Pompeii in 84 CE, to the devastating earthquake of December 2006 from which the people of Pakistani Kashmir are still recovering – but we are powerless to prevent their recurrence. This Tethyan belt eventually merges with the Pacific Ring of Fire, another mighty belt of extreme earthquake and volcanic activity that rims the Pacific Ocean.

The tectonic plates I refer to above are fundamental to the nature and expression of geology at the Earth's surface. Understanding their formation, movement and destruction, therefore, is crucial to charting the expansion and demise of Tethys. A brief explanation of the plate tectonic theory is required at this stage. It is built upon the former static view of Earth as comprising a very hot iron-rich central core, a very thick surrounding mantle made of extremely dense silica-rich rock material, and a relatively thin outer crust made up of the great variety of rocks we see about us at the surface. The plate tectonic concept presents a more dynamic view and a more multi-layered Earth (Fig. 1). The outermost crust is tightly bound to the upper mantle; together they form a relatively cool, and therefore rigid, *lithosphere* (from the Greek *lithos* meaning *stone*). It is this part of the Earth system, 120–180 kilometres thick, that is fractured into a series of distinct (tectonic) plates, which ride on a weaker, hotter, almost molten (or partially molten) layer known as the *asthenosphere* (from the Greek *asthenes* meaning *weak*). The asthenosphere, like the lithosphere, is around 100–200 kilometres thick.

Currently there are 10 very large and some 20 smaller, rigid plates that are in constant motion with respect to one another, as though jostling for position or supremacy. Plate boundaries are irregular and

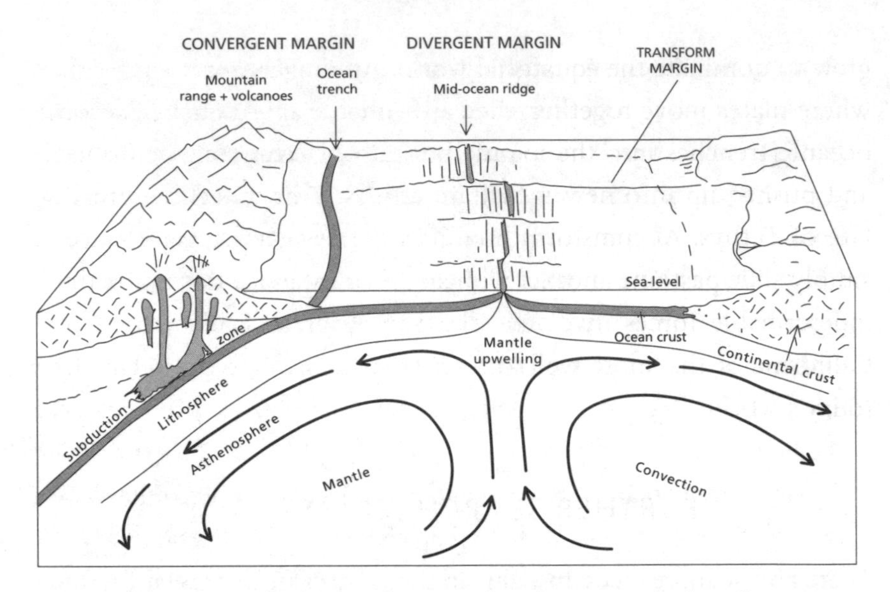

FIG. 1 Plate tectonic schema showing the elements of plate formation, seafloor spreading and subduction.

interlocking, like a giant, spherical jigsaw puzzle. Rates of plate motion are imperceptibly slow to the human time–space frame, being measured in just centimetres per year – about the rate at which our finger nails grow (slow plate movement) or hair lengthens (faster movement). Not only do plates move, but their size and shape constantly changes as new material is continuously being added, and old material consumed. This remarkable metamorphosis takes place at plate boundaries, the giant cracks and healed sutures that scar Earth's outer crust, and along which the majority of tectonic activity occurs – the earthquakes and volcanoes, hot springs and heat loss from the Earth's interior.

Three different types of plate boundary exist (Fig. 1). At divergent plate boundaries new ocean crust, or lithosphere, is continuously being formed. These are known as spreading centres or mid-ocean ridges, where new ocean floor is added and ocean basins grow in size. It was from just such mid-ocean ridges that the Tethys Ocean grew to dominate the equatorial world. At convergent boundaries, where plates move together, they are either dragged back down at oceanic trenches into the mantle beneath or crumpled, deformed and pushed up into new mountain ranges. This was the ultimate fate of Tethys. At transform boundaries, the gigantic, rigid plates simply slide past one another with an apparent grace that belies the indescribable forces involved. The San Andreas Fault system of California is the most well-known transform plate boundary in today's world.

FURTHER SEARCH FOR CLUES

There are far more clues available in the landscape all around us, the mountains and the valleys, and in the rocks they are made of. On close inspection, there is an astonishing array of clues in the fossils contained within some of the rocks and even in the individual grains of sand or lime of which they are made. About 50 kilometres inland from where I am writing, the lesser known Andalucian market town of Antequera borders on the fertile valley plains of the Guadalquivir River and nestles against the sharply defined Torcal mountains to the south. While searching through the weathered grey limestone of these mountains, I found fossil sea urchins (echinoids) and other marine treasures that place the region deep on the Tethys sea-floor around 100 million years ago. The exact species of echinoids and belemnites (fossil squid-like creatures) that I found can be compared with the global fossil database amassed and painstakingly catalogued by palaeontologists over the past 200 years. Using this, we can date the fossils fairly accurately (typically to within a few million years) and, furthermore, determine whereabouts in the ocean they once lived, largely by comparison with where similar creatures are living today.

This reasoning is based on the *principle of uniformitarianism*, a hugely important maxim for geologists, which states that 'the present is the key to the past'. I have just returned from a field excursion to another part of Andalucia, where I was explaining just this concept to second-year university students. We were examining a series of fossilized coral reefs, now stranded high in the hills surrounding Tabernas basin. These *Porites* corals, so the maxim holds, almost certainly lived in warm shallow seas as part of a thriving marine community, just as corals do today. We find other parts of that community also fossilized in the Tabernas reefs – calcareous (red) algae, bryozoans, thick-shelled oysters and gastropods (sea snails or winkles) – many of them were clearly able to withstand constant battering by the waves of that former ocean, and now lie jumbled and tumbled together as part of the reef slope or talus. Others have long since disappeared due to the ravages of time. Intriguingly, these reefs date from the very last throws of the Tethys Ocean about six million years ago, and different phases of reef growth are seen to step down towards the south, chasing the edge of the sea as it slowly receded. Tectonic uplift has since pushed the reefs upwards so that they have assumed their present position flanking the Sierra de los Filabres.

There is an even more remarkable story preserved in the growth structure of these fossil corals and oysters. Corals add a growth layer for each day of warmth and sunlight, thicker layers during summer months and thinner over winter. Oysters of the intertidal zone respond to the tides, adding a thin layer of calcite to their shells on each high tide, and thicker layers during high spring tides. By counting these layers in fossils of different ages and noting the changes in thickness, we can infer the length of the year and the month back through time, for as long as such organisms have inhabited the Earth. Not surprisingly perhaps, we find no perceptible difference in the solar and lunar periods between now and the *end* of Tethys' story, as evidenced by the Tabernas fossils, but at the birth of the Tethys Ocean 260 million years ago, data from other parts of the world have shown that there were around 278 days in a year, 29 days in a lunar month and 23 hours in each day! The earliest fossil data reveal the year had more than 420 days around 600 million years ago, while the lunar month was 30 days long and the day just 22 hours. The most likely explanation for changes in the length of the year and month must be related to planetary orbits – the Earth is moving ever closer to the sun, while the moon is moving slowly away from Earth. Astronomers also calculate that the rotation of the Earth is gradually slowing down as a result of tidal friction.

My own research speciality, within the broader science of geology and oceanography, is deep-sea mud. However, if I give this in answer to the question 'What do you do?', I find that conversation soon wanes. Perhaps a little more glamorously (and when I want a longer discussion) I might describe my work as trying to understand the nature and shape of the deep ocean floor, the life and death dramas it bears witness to, the hidden treasures and useful resources it contains.

The deep oceans of today's world are one of the most remote and hostile, least known and still inaccessible environments on the planet. How much more so, then, the sea-floor of a long vanished ocean. Surprisingly, however, this is not quite true; let us return to the Tabernas basin and I will explain. Standing on the fossilized reefs with my students, we hold a commanding view over a dry and arid landscape, deeply incised with a network of *ramblas*, or dry river valleys – these are the 'badlands' of early spaghetti westerns that were filmed in south-eastern Spain by the famed director Serge Leone rather than Texas. The reefs fringed the last of the Tethys Ocean, with their warm shallow lagoons and frenzy of colourful life. This was still just a few million years too early for scuba diving, of course, although the first hominids had already evolved by this time in the cradle of the East African Rift Valley and some may even have made their way to hunt and play along the southern shoreline of the Tethys Ocean.

Both margins of Tethys became earthquake-prone in these latter stages, as the opposing plates were forced ever closer together. We can see this near Tabernas where great fault scars cut through the landscape and separate mountain from plain. The movement of rocks grinding past each other along such fault zones, repeatedly sticking and then slipping, is what causes earthquakes. The rocks in the fault plane itself become fragmented into a jigsaw-like breccia and then are finely ground or mylonitized, and hence softer and more prone to weathering in a subsequent era. They may also become highly coloured as fluids pass along the plane and precipitate out minerals within the fault gouge, which is exactly what we see in the Tabernas faults – deep reds and purples, even yellows and greens in places. But earthquakes that occur along an ocean margin, beneath the piles of sediment that collect in the shelf and slope areas, have a very different effect. These earthquakes are likely to destabilize whole portions of the slope, which can career downslope at great speeds as huge submarine slides and debris flows move towards the deepest part of the sea-floor. This sort of sudden, catastrophic event is exactly what generates giant tsunami waves in the overlying ocean waters, with all their attendant potential for wreaking environmental havoc.

Standing now in the *Rambla del Inferno* is a seminal moment for me as it would be for any specialist in deep-water sediments. For I am actually standing on what would have been the deep sea-floor of the former Tethys Ocean, now eerily quiet save for the faint rustle of wind and the scurrying scorpion I have unwittingly disturbed. If I had been here 20 million years ago, I would now be covered by 2000 metres of water. But here today I can examine at my leisure and in great detail the nature of sediments that once covered this darkly hidden world. One such deposit is a very thick and chaotic mélange of distinctive grey-coloured rocks that have clearly come from the Sierra de los Filabres mountains some 15 kilometres to the north. The rock clasts of all sizes and shapes, now encased in a hardened mud – sand matrix, have the telltale fabric and glistening mica crystals of a metamorphic rock derived from an ancient continent. Some even contain semi-precious ruby-red garnet crystals, which are only formed at relatively high temperature and pressure in rocks subjected to deep burial and which have most likely been caught up in a period of mountain building where continents collide. Today such garnets are used locally for jewellery and even as an abrasive in sandpaper.

This distinctive deposit is as much as 70 metres thick in places, the same as my whole class of some 40 students standing on each other's shoulders, and would have once covered the whole of the Tabernas basin, before the effects of badlands erosion – an area of around 400–500 square kilometres. It is charmingly referred to by geologists as *El Gordo*, meaning the 'big one' with reference to the Spanish national lottery. We know from its particular characteristics that it was deposited as a gigantic submarine debris flow, by just the same process as that observed on land in mountainous areas after torrential rainfall or rapid snow melt, all too common and often with devastating the results for anyone living in its path. An earthquake somewhere on the slope in the vicinity of the Tabernas reefs triggered a massive submarine slide, which became more and more unstable and chaotic as it avalanched down the steep submarine slope, ingesting copious volumes of seawater as it flowed. The powerful flow then spread its lethal load over the entire Tabernas basin – a once quiet inlet or embayment from the Tethys Ocean.

We know it came originally from near the reef because *El Gordo* also contains jumbled blocks of coral debris, some of which have polished up well and furnish most attractive bookends in my study at home! We also know that it must have been lethal, for nothing once living on the sea-floor could possibly survive sudden burial beneath such a thickness of mud and rubble.

But the fascinating story of devastation does not end yet – there are still other clues to unravel. Overlying *El Gordo*, everywhere I looked was a much thinner layer (about one metre thick) of fine sand grading up into silt and mud. This kind of deposit is something that geologists call a turbidite, as it is deposited by a turbulent suspension, or turbidity current, as it flows across the sea-floor like a submarine river in flood. But why is it there and what might it signify? We know from countless modern events that the sudden displacement of a vary large volume of material either into or within the sea is likely to generate a tsunami. In this semi-enclosed arm of the Tethys, the Tabernas tsunami of some 7.5 million years ago would have rampaged up the opposing landmass, now represented by the Sierra Alhamilla, and then fallen back towards the sea, carrying with it a swirling load of beach sand and shelf mud. This potent mixture formed the turbidity current that blanketed its load over the top of *El Gordo*. This last may be a more fanciful explanation of the data than is necessary, but it is by no means unrealistic.

Returning now to the mountains themselves. From those that surround Tabernas, through the limestones of Torcal National Park and the Sierra Nevada, and in a great arcing sweep across southern Spain all the way to Gibraltar, these are all part of the Betic mountain ranges, formed when the African and European plates finally rammed together in this part of the world and slowly squeezed the Tethys Ocean out of existence. All manner of other clues are there to be unravelled, some of which are discussed in later chapters, but I must mention just one more important one at this stage. That is the Ronda *ophiolite* – an exhumed portion of ocean crust. In fact, it was my wife, Claire, who first noticed the distinctive shiny and greenish-coloured rocks by the roadside as we wound our way inland one simmeringly hot summer's day. She had recognized them as very similar to those of the Troodos mountains in Cyprus, about which I had much enthused on a previous excursion.

Ronda is another charming hilltop town within easy striking distance of the *Instituto Espanol de Oceanografia* here in Fuengirola, and all around it in the mountains are rocks of unique significance to the Tethys story. They are hard, compact and surprisingly heavy rocks, dark grey-black in colour when fresh but more commonly weathering to a magnificent shiny green. Massive quarries scar the landscape, for they are an immensely tough and durable rock when used as road metal, as well as being a most attractive ornamental stone for building works. To the trained eye they are immediately recognizable as what geologists refer to as ultrabasic rocks, and that go by wholly unpronounceable names such as peridotite, harzbergite and lherzolite, or serpentinite for the greenish variety. To be certain of our identification we have to take rock samples back to the laboratory for thin-sectioning, which involves cutting them into thin slivers with a diamond saw, sticking these rock slivers onto a glass slide with strong resin and then polishing them down until they are exactly 30 microns in thickness. That's about a tenth the thickness of an average grain of sand, so thin that light from a petrographic microscope can shine directly through the rock and allow us to identify the individual minerals present.

Sure enough, thin sections from the Ronda samples show the suite of minerals characteristic of ultrabasic rocks, especially olivine and serpentinites. Such rocks are quite exotic to the outer crust. They form at much greater depths in the Earth, in the underlying layer known as the mantle. Almost all the examples we know of have been squeezed up from the mantle that underlies the ocean floor, often together with the volcanic ocean crust and overlying deep-sea sediments in a complex known as an ophiolite. The greenish serpentinite forms when seawater interacts with chemically unstable ultrabasic rocks as they are pushed upwards. The evidence, therefore, is conclusive – the Ronda ophiolite has been exhumed from the very bowels of the Tethys Ocean, in this case having formed when the ocean was almost at its maximum extent some 100 million years ago.

THE SLOW BEAT OF TIME

Geologists are obsessed with time and yet, as my wife has noted in my case, we are quite hopeless timekeepers! It is perhaps because we understand the meaning of geological time, or *deep time* as it is sometimes called, which is measured in millions if not billions of years and is always imprecise, that we may sometimes be a few tens of minutes late for an earthly appointment. But let me now explain something of deep time and just how it is measured. How is it possible for me to state so adamantly that the Tethys Ocean reigned supreme between 260 and 5½ million years ago, that is, for a quarter of a billion years of Earth history? How do we know that this period of history includes the entire 'Age of the Dinosaur' and almost the whole 'Age of the Mammal' right up until the point that our distant ancestors began to walk upright and, perhaps, to distinguish themselves in other ways from our chimpanzee cousins?

Two aspects of time are significant here, both the *relative* sequence of events in the rich kaleidoscope of ocean and earth history that we seek to chart, and the *absolute* age of rocks, fossils, past ice ages, and almost everything else we observe. This includes the age and duration of the Tethys Ocean, a matter I shall return to shortly. Once the relative and absolute ages are known, it becomes possible to determine the rates at which natural processes operate – from the inexorably slow movement of plates, the building and denudation of mountain ranges, to the speed with which earthquake tremors pass around the globe and through the Earth's interior, the sequence of waves travelling across an ocean basin, or the rhythm of the tides. We are therefore equally obsessed by the measurement of time and have devised a series of different timescales (or *stratigraphies*) that help in our quest to describe, understand and measure natural earth phenomena.

Recognition of the *Law of Superposition* came early in the history of geology. This holds that sedimentary layers or strata, such as my deep-sea mud, sands or the limey ooze that later turns to chalk, are laid down sequentially on top of one another so that in a normal succession of sedimentary rocks (i.e. one that has not been overturned by subsequent earth movements), the oldest rocks will lie at the base and the younger ones at the top of the pile. A *litho*stratigraphic (*rock* stratigraphic) timescale can thus be constructed for each sedimentary rock succession exposed on land or in boreholes drilled into the sea-floor, which allows us to read through the sequence of layers, with their variously encoded inscriptions of Earth history, almost like leafing through the pages of a book. If we read from base to top of the pile, then the relative history of events is revealed for that particular region. For example, the sequence of rocks below the Tabernas reefs, which I referred to earlier, show unmistakable signs of a shallowing water trend up to the capping reef, supporting the concept of the progressively retreating Tethys Ocean.

The systematic record of evolutionary changes in the Earth's animal and plant life, drawn up through painstaking research on fossils by palaeontologists, led to the development, and now constant

LIBRARY
THE UNIVERSITY OF TEXAS
AT BROWNSVILLE
Brownsville, TX 78520-4991

refinement, of a biostratigraphic timescale. This has the added advantage of being more readily correlated from one part of the globe to another, especially for those fossils and microfossils that spread right across the globe. It was these subtle and systematic changes observed in fossils through sequences of rock layers in different parts of the world that provided important contributory evidence for Charles Darwin in developing his revolutionary new theory on the origin of species through natural selection, which now, of course, underpins the whole of evolutionary theory. It was by reference to this biostratigraphic timescale that I could so confidently place the Torcal fossils near Antequera to their position within the overall history of the Tethys.

One of the most recent and carefully refined among these relative timescales has been derived from recognition of cyclic changes in Earth's climate and of the effects that these have had on a whole variety of natural events. These are known as Milankovitch cycles after the Serbian mathematician, Milutin Milankovitch, who so elegantly demonstrated that their periodicity (on timescales of 100,000, 41,000 and 22,000 years), corresponded with astronomical changes in the Earth's orbit, tilt and axial wobble, respectively. Variations in global sea level, the extent of ice cover and even in the amounts of ice-rafted debris incorporated into sediments from melting icebergs, can all be linked back to this astronomically forced climate change. In the same way, variation in mean temperature at the Earth's surface induces very subtle changes in the ratios of chemical isotopes (principally oxygen and carbon) that have been incorporated into the shells of living organisms as they grow. Sophisticated instruments now permit the routine measurement of these chemical ratios in the record of microfossil shells taken from vertical cores of sediment recovered from the sea-floor. A whole new stable isotope stratigraphic timescale has thus been born.

There are other timescales that I have not yet mentioned. For example, periodic reversals in the direction of the Earth's magnetic field can be measured in the magnetic properties of rocks and used to build up a magnetostratigraphic timescale. This enormously important technique will be returned to in a later chapter when looking at how we can chart an ocean's growth from its birth to eventual demise. As science constantly seeks to refine its measurement of time, then inevitably new timescales will be developed in the future.

However, none of these techniques enables us to assign absolute ages, to state with confidence that the Earth formed 4.5 billion years ago, that the oldest ocean crust is 180 million years old, that the last dinosaurs roamed the world 65 million years ago, that modern humans first appeared 4 million years ago, and so on. This was still the position of our science towards the end of the 19th century, despite the fact that geologists had been steadily pushing back the frontiers of time from the earlier religious contention that the world was created in 4004 BCE. A much longer time frame was needed to allow for Darwinian evolution, for the building and destruction of mountains, and for the salting of the seas. A means of absolute dating was desperately needed.

This possibility came with the discovery of radioactivity by the French scientist Antoine-Henri Becquerel in 1896, and then in the early 20th century with the application of radioactive decay of naturally occurring elements in rocks to obtain their absolute ages. This principle is based on the discovery that some elements, or particular forms of an element known as isotopes, are unstable and spontaneously transform themselves into a different, more stable isotopes, emitting radiation in the process. Uranium-235 decays to thorium-231 and then, through a series of decays eventually becomes lead-207, while thorium-232 goes through a different series of decays to produce lead-208, and so on. The number after the name of the element,

GEOLOGICAL TIMESCALE and TETHYS OCEAN TIMELINE

ERA	PERIOD (Epoch)	DATE (Mya)	PRINCIPAL EVENTS Earth, Ocean and Biosphere
CENOZOIC	QUATERNARY Holocene Pleistocene	0–2.6	Modern plate tectonics, oceans and continents Modern humans, rodents and insects dominate Ice Age maximum; global conveyor-belt circulation
	NEOGENE Pliocene Miocene	2.6–23	*Tethys* closure complete (5.3 Mya); evaporite crisis India collides with Asia; Panama gateway closes First hominids; modern life evolves Antarctic isolated; icehouse conditions begin
	PALEOGENE Oligocene Eocene Paleocene	23–65	*Tethys* narrows; Atlantic grows; India drifts north Alpine mountains of Europe develop First whales; modern coral reefs proliferate Radiation of life at the beginning of a new Era
MESOZOIC	CRETACEOUS	65–145	*Tethys* reaches maximum extent; Atlantic opens Deccan Traps (India) volcanic superplume Calcareous oceanic plankton excess; Black Death KT Mass Extinction (65 Mya) Highest sea-level; warmest global temperature
	JURASSIC	145–205	*Tethys* cuts through Pangaean rifts, and expands East–West ocean circulation established Modern fishes, marine reptiles, ammonites, and oceanic plankton gardens
	TRIASSIC	205–250	*Tethys* lies to East of Pangaean supercontinent Global rifting episode fractures Pangaea Corals, ammonites, reptiles, first dinosaurs End Triassic Mass Extinction (205 Mya)
PALEOZOIC	PERMIAN	250–299	*Tethys* comes into being as Pangaean supercontinent formation completed (256 Mya); Panthalassa ocean Great Pangaean Mountains across supercontinent PT Mass Extinction (250 Mya) ends Paleozoic life
	CARBONIFEROUS	299–359	Gondwana and Laurasia continents drifting together Major forests; winged insects evolve; ancient corals Caaboniferous-Permian Ice Age grips Gondwana
	DEVONIAN	359–416	Final closure of Iapetus Ocean Old Red Sandstone Continent (Laurentia-Baltica) Ancient fishes, amphibians, large land plants
	SILURIAN	416–444	Baltica, Avalonia and Laurentia collide Caledonian mountain building Vascular plants, early land animals, jawed fishes
	ORDOVICIAN	444–488	Iapetus Ocean dominant low-latitude ocean Trilobites, graptolites, ancient corals, jawless fishes Hirnantian Ice Age
	CAMBRIAN	488–542	Rodinia fragments; scattered continents at low latitudes Very high global sea level Cambrian radiation of life; enhanced fossil preservation
CRYPTOZOIC	PROTEROZOIC	542–2,500	Rodinia supercontinent forms Multicellular organisms evolve, still entirely ocean-bound Ediacara soft-bodied fauna proliferate (640 Mya)
	ARCHEAN	2,500–4,500	Origin of Earth (4,600 Mya) Core–mantle–crust structure forms Differentiation into ocean crust and continental crust Oceans form (around 4,000 Mya) Life originates on Earth (around 3,500 Mya)

duration of the Tethys Ocean

FIG. 2 Geological timescale with geological eras and periods. Some of the principal geological events are listed, especially as they affect the Tethys Ocean. Mya = Million years ago.

by the way, refers to the total number of protons and neutrons, the subatomic particles that together make up the atomic nucleus.

What is so remarkable and fortuitous about this process of decay by radioactive emission is that each radioactive isotope decays to its stable 'daughter' isotope at a rate which is unique to that element. This provides us with a series of radiometric clocks, for some of which the ticking of the clock is extremely slow, while for others it is geologically rapid. For example, half the uranium-235 in any given rock will change to lead-207 in 713 million years (this is known as its *half-life*), whereas the half-life for carbon-14 is just 5730 years. The former can be used to date the oldest rocks on Earth, the latter to date organic material no older than 75,000 years.

So, finally, by selecting the right isotopes naturally occurring in rocks or fossils for dating, we have a means of calibrating the whole of the relative stratigraphic timescale, and all its variations – magnetostratigraphy, isotope stratigraphy, and so on. This discovery heralded a major breakthrough in science, a quieter revolution than that of plate tectonics, perhaps, but a very crucial one nonetheless. The successive layers of rocks now provided a true stratigraphic *timescale*; the eras, periods and epochs were assigned absolute ages of which we could be relatively confident. Almost every geologist I know carries a pocket-size version of this timescale wherever they travel, a handy reminder of the principal chapters in Earth's incredibly long history. Ask the next geologist you meet and they will doubtless produce it from between their credit cards and driver's licence, or from just in front of the latest family photographs!

A version of this stratigraphic timescale is provided in Fig. 2, with the addition of some of the principal geological and oceanographic events that have shaped the world we know today. This also shows the duration and stages that mark the story of the Tethys Ocean and hence signpost the chapters of this book. But do not be deceived.

This is a portrait of the whole of Earth history in just one greatly simplified diagram. The timescale involved is mind-bogglingly long; it is the awareness of deep time that contributes to a geologist's hopeless sense of earthly time but also, I believe, to a more realistic sense of our own place in deep history and of the inevitable transience of all things – of oceans, mountains and especially of individual species.

2

Pangaea the Supercontinent and the Birth of Tethys

Darkly hidden, lost in time, the age-wrinkled
marble mountains, remnants of bygone eons,
thrust upwards from a vanished ocean's mighty depths,
a sheer limestone face blushes to the evening sun.

from *Filomena Muse* by Dorrik Stow

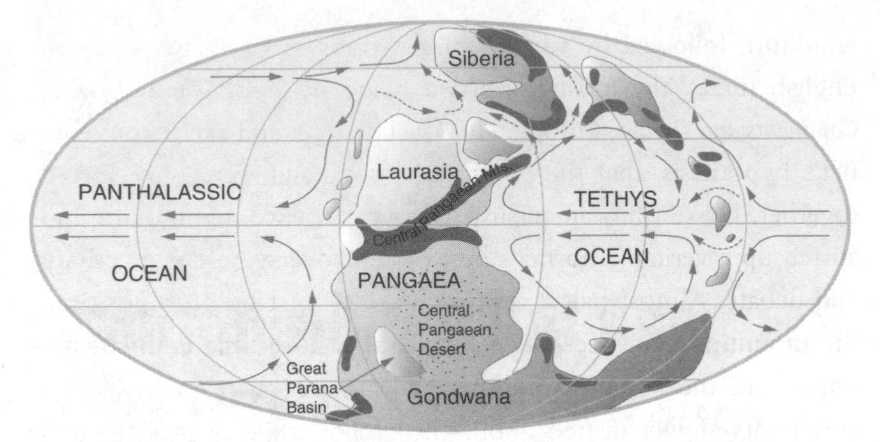

Late Permian Tethys map (260 Mya – million years ago). Global reconstruction with oceanic circulation, also showing principal mountains, deserts and evaporite deposits.

In a time before even dinosaurs had evolved, around 260 million years ago (260 Mya), the world was a wholly different place. The continental plates that make up the lighter, more buoyant part of Earth's outer crust, had become fused together into a single supercontinent, known as Pangaea. This stretched from pole to pole, surrounded on almost all sides by a great global ocean known as Panthalassa. A giant C-shaped bite on its eastern margin is our first glimpse of the Tethys, a tropical ocean straddling the equator and separated from Panthalassa by an archipelago of islands, including what is now China and South East Asia. At the heart of Pangaea lay perhaps the greatest mountain range of all time, from which monstrous rivers drained directly into the Tethys, and a fearsome red desert, larger and hotter than anything the world has since known.

The concept of this global supercontinent was originally put forward in 1912 by a young German meteorologist named Alfred Wegener in his address to the local Geological Association of

Frankfurt, followed by various publications in German (1915) and English (1924), the most important being his book *The Origins of Continents and Oceans*. Completely novel, elegant and exciting, Wegener's hypothesis, that the continents move and were once linked together, broke onto an unsuspecting geoscience community and stirred up a veritable hornet's nest of controversy, research activity and debate. Almost half a century after his first bold proposition, the mounting evidence had become incontrovertible and, equally important, the scientific rudiments of the mechanism by which continents slowly drifted about the globe, while oceans opened and closed before them, were understood. The controversy resolved into a fundamentally new understanding of planet Earth, the processes that operate deep within as well as those that shape its surface, the nature of mountains and oceans, the hidden meaning of earthquakes and volcanoes, and of much else besides.

From Thomas Kuhn's 1970 book, *The Structure of Scientific Revolutions,* we can take this 20th century upheaval in the geosciences as a classic example of how normal science evolves slowly and safely within one dominant paradigm and then, through a period of rapid and tumultuous change, adopts a completely new paradigm. For the geosciences, this particular scientific revolution culminated in the early 1960s and the new world view, accepted by almost all earth and ocean scientists today, became known as the *plate tectonic* paradigm. The names of various individual scientists loom large according to how significant their contributions were to this evolving paradigm – largely dependent on which side of the Atlantic is writing the history, or on whether their perspective is primarily geological or oceanographic. Assigning that honour matters little to plate tectonics, of course, and the reality is that the many elements of our current understanding were championed by different people.

It is a sad quirk of fate that Wegener himself never lived to witness the profound repercussions of his bold new theory. He died in 1930 at the age of just 50 while on a German expedition to conduct the first 12-month monitoring of Arctic weather. He died from exhaustion and exposure, stranded high on the middle of the two-mile-thick ice sheet that covers Greenland, caught as the most extreme weather conditions conspired against his final bid to reach the expedition base camp at Umanak Bay on the coast of Baffin Bay. He was an explorer and scientist of exceptional insight and talent, whose name is indelibly linked to the theory of continental drift and hence to the start of a major scientific revolution. I always find it interesting that many of the major breakthroughs in science (and indeed in other disciplines too) are made by those with creative and enquiring minds willing to trespass outside their own 'comfort zone' of special expertise. Wegener's trespassing from meteorology into geology was not welcomed by established geologists at the time, but he is now heralded as the grandfather of plate tectonics.

Knowing that the system of plate tectonics operates is one thing, but actually reconstructing the nature and position of past continents and oceans is another – an extremely complex affair. It also becomes vastly more difficult the further back in time we dare to travel. For an insight into such matters of the deeper past I have always consulted two of my former colleagues from the University of Nottingham, Drs Clive Boulter and Tim Brewer. They were both enthusiastic young lecturers with me at Nottingham and have taught me much over the years. Sadly, Tim passed away as I was beginning to write this book, but I'm sure that he would not mind me recalling his reference to my 'tinkering with the Tethys' as 'merely gardening'. This was in comparison to his own work on very ancient Precambrian continental landmasses (also known as cratons where they are relatively undeformed), such as that of Greenland/Canada on which Wegener met his untimely end.

As far as we know from work on these Precambrian cratons, plate tectonics started to operate some time during the Archaean Eon, a period that covers the first half of Earth history up to 2.5 billion years ago. Certainly by the beginning of the Proterozoic Eon that followed, normal plate tectonic processes were under way, the cratons were moving about the surface of the Earth, and new ocean crust was being formed at mid-ocean ridges. These cratons were also growing steadily in size by a process of repeated collision and mountain building, and the addition of such newly formed mountain ranges to the edges of the continents. Geologists of the Precambrian have gathered sufficient evidence to propose that around 1 billion years ago, maybe a little less, there was an earlier supercontinent which they have called Rodinia. But very little is known of its assembly or demise, all of which took place long before the much better known and even larger supercontinent, Pangaea. It was against Pangaean shores that waters of the Tethys Ocean first lapped but a quarter of a billion years ago – the realm of gardening!

I want to first outline in this chapter just how the Pangaean supercontinent came into being and then consider more briefly the early Tethys Ocean.

MOUNTAIN AND GLACIER

Construction of the largest single continent the world has known was no inconsiderable feat. The slow but inexorable juggling together of separate continental plates took at least 50 million years, the forces involved in their ultimate fusion were immense, and the scars across our global landscape still remain to be deciphered today. Essentially it is a story of mountain building (orogeny), for when plates collide then the rocks between are squeezed upwards into mountain ranges. These processes operate very slowly and intermittently, always accompanied by the chaos and devastation wrought by earthquakes and volcanic activity. The fastest rates of uplift in mountainous areas today, in parts of the South American Andes for example, are only around one centimetre per year, which translates to some ten metres per thousand years. A more average rate of active mountain building is about a tenth of this value, or around one metre per thousand years. Interestingly, even this average rate of uplift, operating over one million years, would potentially yield mountains 10,000 metres high, towering high above Mount Everest and the other mighty peaks of the Himalayan range. As most orogenic episodes last for *many* millions of years, including the latest one that produced the Himalayas, there is something of a conundrum here – simple multiplication doesn't work.

There was already one very large continent in the southern hemisphere, known as Gondwana, which had been in existence for much of the Palaeozoic Era. This incorporated what have now become the widely separated landmasses of South America, Africa, India, Australia and Antarctica, and had been known or suspected by geologists for some time, based on the large number of identical species of fossil plants and animals common to these diverse regions of the world. This was true, for example, of the seed fern *Glossopteris*, an important plant typical of the Permian Period and used by Alfred Wegener in support of his theory of continental drift, whereas more conventional paleontologists at the time believed that the seeds must have blown across the oceans to fertilize each continent separately.

After suture of its several component parts, Gondwana moved around the southern hemisphere as a single coherent plate. Intriguingly, the history of this movement can be charted, in part, by observing patterns of past glaciations recorded in rocks of different ages in Palaeozoic rocks of this southern land that are far, far older than the tell-tale signatures left by the receding ice of the most recent glacial epoch. The clues, however, are much the same for glaciations of both the recent and distant past.

I have sought out this evidence of a late Palaeozoic glaciation on all the disparate continents that were once joined as Gondwana (South America, Africa, India and Australia) – all except for Antarctica, that is, where the clues are mostly buried beneath the present-day ice cap. In particular, I well remember a clear, bright day one austral winter while out walking along a cliff-top footpath near Adelaide in South Australia. A chill wind blew into our faces as the waves spewed foam across the dark rocks below, with nothing but a broad and wild expanse of the Southern Ocean between wind, spray and the Antarctic continent from whence these sports of nature had originated some days before.

Now, I am always intrigued by the harsh climatic extremes on our little planet – from the incomparable cold and isolation of central Antarctica to the inhospitable inferno of the Great Australian Desert or Saharan sand seas. Warming to my theme, I was explaining to my youngest daughter, Kiah, how these extremes only occurred when the world was in an icehouse phase with polar icecaps as it is now, notwithstanding human-induced global warming. The last time this occurred was over 250 Mya when Australia was joined to Antarctica, and this part of South Australia nestled up against Wilkesland. 'But how do you know, Daddy?' was Kiah's simple and very direct response, always an effective challenge in this sort of conversation.

Musing a little before I replied, we rounded a headland and walked right across a broad flat surface of polished rock. I stopped dead, completely lost in awed silence, staring past my feet in wonder at the criss-cross pattern of glacial striae and pluck marks that adorned the smoothed surface – exactly the same markings as are created by glaciers today or by those of the most recent ice age as they scrape and grind the rocks below. But the rocks polished and sculpted beneath our feet were *Permian* in age, scoured by Permian glaciers. Here was my evidence, in fact very helpfully confirmed for my family by one of those rare scientific explanation boards put in place by the Australian Parks Authority.

Exactly this sort of evidence, together with a characteristic type of glacial deposit – a chaotic mix of mud, sand and boulders known as 'boulder clay' or 'diamictite' – which is found scattered across the southern continents today, is what leads us to conclude that the last great icehouse phase gripped planet Earth in the late Palaeozoic Era when Gondwana was drifting ever so slowly across the South Pole. We can trace that polar wander from South America, through southern Africa and thence to Antarctica. Rather surprisingly, Antarctica has remained close to or directly over the South Pole ever since the early Permian Period, while other parts of Gondwanaland became detached and drifted away.

The plate movements that brought the southern part of this great landmass near to and then over the polar region were driving the other parts of Gondwana ever northwards into full plate collision with the northern continent known as Laurasia. For at least 50 million years this monstrous duel ensued. There were no winners or losers, save for the rumpled edges of the giant continental plates. A former ocean that stretched between the two for more than 1000 kilometres was slowly consumed, and as the cold oceanic slab descended deep into the bowels of the Earth it melted and threw up several strings of volcanic islands. These archipelagos were also summarily destroyed as great thicknesses of lava flow and ash-fall, caught up with fringing coral reefs and terrestrial sediments derived from the advancing continents, were squeezed together under enormous pressure, and then folded and pushed upwards into a new mountain range.

As far as we know, admittedly from the limited evidence that remains from the earliest periods of Earth history, the world had never previously witnessed plate collision of quite such longevity and scale. The mountain building episode that finally fused together the greater part of the Pangaean supercontinent goes by several names, partly because it covered such a large area and partly because of its evolution through time. The Hercynian, Acadian, Appalachian, Alleghenian and Ouachitan orogenies have each lent their name to a part of this greatly extended episode. Taking a simple view of this complexity, we can say that collision began in the east, in what is now part of North Africa, and ended in the far west, in what are now the south-western states of North America. Again for the sake of simplicity, I will call the sum of these events the Great Pangaean orogeny, and the full sweep of mountains that resulted has already been dubbed the Central Pangaean Mountains by others before me. This colossal range stretched about 7000 kilometres, clear across the supercontinent, and was probably as much as 1000 kilometres wide in places. But how high did the most majestic peaks reach into those primeval skies, and were they snow-capped or thickly vegetated?

As I mentioned earlier, we cannot estimate their original height simply by multiplying an average rate of uplift (say one metre per thousand years) by the 50-million-year duration (at least) of the Great Pangaean orogeny. This would yield mountains 50 kilometres high! As soon as uplift begins then erosion follows, and the higher the mountains grow the greater the rate at which they are worn down. This is the result of two main effects – first, higher and steeper peaks are naturally prone to greater mass wasting by rock fall, landslides and debris flows; and second, the ice and snow that form at high elevation are especially effective agents of weathering. In fact, the pace of erosion soon outstrips that of uplift, such that over large areas of today's mountain ranges, the Alps or Andes for example, an average of 1–3 metres of rock is being removed every thousand years. In parts of the High Himalayas this can be as much as 15 metres per millennium.

There is a further constraint on the ultimate height of mountains that is due to the huge excess weight that their sheer bulk generates for the continental plate on which they reside. The plate responds by something akin to elastic flexure, which creates a tendency towards subsidence rather than uplift. Controlled by these dual limitations, that of increased erosion and regional flexure, it seems unlikely that the highest of the Central Pangaean peaks exceeded about 10,000 metres at their zenith. Nevertheless, they would have dwarfed Mount Everest and K2, their rarefied atmosphere would have supported no life, and the snow-capped line of these imposing peaks most likely stretched from coast to coast. This is despite the fact that they were in very much an equatorial to subtropical range, such that their lower slopes were clad with a rich array of Permian plants, and these primeval forests were home to a veritable kaleidoscope of long-lost fauna.

Pangaea was as complete now as it would ever be. Some geologists contend that it was really a close assembly of continents, rather than a single supercontinent. But I prefer the latter interpretation, especially having trekked across many parts of the former Great Pangaean Range and seen for myself the evidence of that supreme suture between Gondwana and Laurasia. I have also visited the Ural Mountains, where Kazakhstan and Siberia finally fused into Laurasia and so closed the late Permian Ural Sea. There were other ranges too, where Antarctica abutted South America and Australia.

DAWN OF THE TETHYS

For me this is also the geological period at which to clearly designate the large C-shaped ocean that straddled the equator east of Pangaea as the Tethys Ocean. It is worth noting that some authors write of a Proto-Tethys, which existed previously, and both Palaeo-Tethys and Neo-Tethys Oceans, which have come since. I simply use Tethys for the ocean that appeared 260 million years ago when Pangaea was complete. I have calculated the size of Tethys at this time at between 80 and 100 million square kilometres, larger than the Indian Ocean and probably a little larger than the Atlantic. There were some outlying land areas – small continents and island archipelagos – that partly separated Tethys from the rest of the Panthalassa Ocean. These islands of the eastern Tethys were probably much like those of the western Pacific today – a mixture of tropical paradise, dangerous volcanic seamounts rimmed by extensive reefs, and larger mountainous lands covered in dense vegetation. Much later in the course of Tethys history these were to fuse into Laurasia and so form the bulk of what is now China and South East Asia.

We can still find glimpses of the strange marine fauna from this period preserved in rocks now pushed up onto land at what were the western and eastern extremities of Tethys. I have come across many relatively large microfossils while working in Sicily, Crete and on the Greek island of Hydra, including cigar-shaped fusulinids and the tiny seed shrimps, known as ostracods. Further to the east, at Tamba in south-western Japan, Guizhou and Guangxi in southern China, and Selong in Tibet, there have been unique finds of brachiopods (lamp shells), coiled goniatites (of nautiloid affinity), and conodonts. All the species discovered are now long extinct, but the conodonts are especially puzzling. These are tiny phosphatic tooth-like structures, about the size of a grain of sand, which occur in great abundance in most Palaeozoic sediments, and they finally disappeared at the end of the Triassic. They occur in a fantastic variety of types, evolving rapidly with time. For more than 100 years they have been carefully studied and much used by paleontologists for the dating and correlation of rocks around the world. But no one had the slightest idea what they actually were, nor what animal they came from – until a chance breakthrough in the early 1980s.

Another colleague from my Nottingham days, Professor Dick Aldridge (now at Leicester University), was working with Derek Briggs and Euan Clarkson of Edinburgh University on some Palaeozoic mudrocks from Scotland. Although the rocks had been collected some time previously and already archived in a museum, no one had noticed the fossil impressions of minnow-sized elongate organisms with conodont fossils lying in situ as a complex feeding apparatus. Finally, the conodont animal or animals had been found. They were tiny eel-like creatures, just a few centimetres long, with two large eyes and chevron-shaped markings along the body indicative of muscle attachments found only in chordates. These very primitive vertebrates had a very long and hugely successful history. Different types appear to have adapted to all marine environments, from shallow to deep water, from equatorial to high latitudes, and to have lived in their millions throughout the early Tethys.

HEART OF THE CONTINENT

But back now to chart some more of the assembly and nature of Pangaea, I will travel to Morocco and descend from the dizzy heights of the Central Pangaean Mountains to the heart of this ancient continent. For all too soon the world would change once more, the supercontinent would fragment, and prehistoric life continue to evolve, such that neither would ever be the same again. Fortunately, however, the remnants of this land of such colossal magnitude still abound in North Africa.

Marrakech is one of my favourite cities and has several times afforded me the opportunity to venture further into the Pangaean past. It is a Berber city founded in 1062 by warrior monks from the Sahara, the Almoravids, whose empire and influence soon stretched from Mauritania to Spain, and eastwards beyond Algiers. They raised ramparts around the city, installed an ingenious system of underground irrigation canals, and even hired craftsmen from Andalusia to build a grand palace and mosque. Although the colour and hubbub of everyday life in Marrakech may have ebbed and flowed through the ensuing millennium, its strength and character endure. The vibrancy and apparent chaos, the noisy bartering for myriad goods, the diligent craftsmen and pungent aromas that so epitomize Arab souks, sprawl through a maze of narrow streets north of the main square. After a day's writing in the peace and tranquillity of La Mamounia Hotel, I venture out to enjoy the contrast of the large, bustling *Place Jemaa el-Fna,* and am soaking in the early evening warmth and atmosphere, sipping a refreshing mint tea and jostling for a place at one the numerous open-air stalls for a spicy lamb tagine with couscous.

Lying on fertile plains at the edge of the Western Meseta in Morocco, and nestling close against the High Atlas ranges, Marrakech is superbly placed for delving further into our Pangaean story. For, at their core, the Atlas mountains are remnants of the Central Pangaean Range as it swept towards its eastern extremity. The main inland road to the north, between Marrakech, Azrou and Fes, and ultimately on towards the Mediterranean coast and Algeria, is as ancient as it has been well travelled through countless centuries. It first skirts the highest peaks and then courses across the Middle Atlas, a remote and isolated region of central Morocco not only preserving proud Berber traditions but also containing a fantastic geology. It was here that I pushed through the dawn market of a hundred white tents on the outskirts of Mrirt and across the hills beyond to find a new geological window on the past, taking me much further back into Earth history.

There are large outcrops of Palaeozoic rock exposed in this region, each between tens and hundreds of square kilometres in area and of different ages, juxtaposed one alongside another. Some show signs of intense metamorphism, having been buried to great depths and subjected to the high pressure and temperature regime found at the core of a major mountain range. Some were less deeply buried and only slightly metamorphosed. At the greatest depths below the mountain belt and subjected to even more intense physical extremes, the rocks themselves become molten – typically at depths in excess of 25 kilometres. The molten rock occupies huge magma chambers where it undergoes a process of fractionation, in which the more dense minerals sink while the lighter fraction rises. This portion of the magma intrudes into the overlying rocks before it slowly – painstakingly slowly – cools and crystallizes into a new suite of minerals. The result is granite, the most common rock of the Earth's *continental* crust, composed of just three principal minerals: glassy quartz, white or pink feldspar, and sparkling silvery plates of mica. The larger bodies of granite are known as plutons, after the Greek god of the underworld, *Ploutonios*; and, indeed, the whole family of rocks formed at great depths from molten magma are termed *plutonic*.

The granites that now outcrop in the Middle Atlas, cut through in places by the P24 highway along which I was travelling, were formed in this way at the core of the Central Pangaean Range. Where the surrounding limestone has come into contact with the intense heat of an intruding granite pluton, it is metamorphosed to marble.

The many varieties of marble, with their intricate swirls and blemishes of different colour, are produced by a host of different mineral impurities such as iron, manganese, copper and cobalt, either from within the original limestone or that seep into the limestone from the cooling granite. It is these *Pangaean* granites and their associated marbles, prized building stones through many ages, that now adorn palaces and mosques, hotels and plazas, railway stations and airports throughout Morocco.

The same marbles were first used by the Roman builders in their reconstruction and extensive development of the small town of Volubilis, following annexation of the whole region, then known as Mauretania, by Emperor Claudius in 45 CE. Volubilis was duly raised to the status of a *municipia* (free town), the largest such conurbation at the far western edge of Roman domination. Its striking remains today lie just an hour's drive north from where I was working near Mrirt in the Middle Atlas, so I was able to visit on several occasions, to admire the fine craftsmanship as well as the judicious use of natural stone. One complete side of the Basilica with its nine magnificent arches remains wholly intact, the colour and detail of *Diana and the Bathing Nymphs* mosaic in the House of Venus is truly exquisite, but for me the Triumphal Arch is the true *pièce de resistance.* It stands astride the column-lined main street, *decumanus maximus,* halfway between the eastern (Tangier) and western gates.

Through this arch to the west there lie fertile plains of cereals and olive trees, much as they must have lain some two millennia before, with vineyards too on the low slopes beyond. As the sun began to set on one particularly fine evening, I could almost glimpse across the open fields flickering torchlight and open fires, and hear some light commotion across the passage of years from dusty Roman legions settling down to their nightly meal. Fanning my imagination that evening, was the knowledge that a former Roman military encampment had recently been discovered by archaeologists from the University of Rabat. It was located just across the western plain from the town, and new excavations were now in the planning. But, dreams aside of this former Roman occupation, let me return to the building materials they used, which was where I started out with my digression to Volubilis. In fact, much of the stone quarried for the Triumphal Arch, as for the rest of the town, was a local sandstone. Marble was used only for certain more ornate parts, for richer houses and for the public baths. Granite was not a preferred building stone at this time because its extreme hardness made it far more difficult to work.

The pale buff and yellowish sandstones, however, were ideal building material – durable but easily worked. In geological terms, they come rather later, following considerable erosion of the Central Pangaean mountains and rise of the Tethys sea level. The sandstone was laid down on the shallow sea-floor fringing the continent, and contains telltale fossils of the Mesozoic Era that was to follow.

TORRENTIAL RIVERS AND SCORCHING DESERTS

There is much evidence also for the first phase of erosion from this imposing mountain range that straddled the Pangaean equator in the late Permian. When I first drove through Khenifra on my way to Mrirt, I was immediately struck by the carmine colouration of so many buildings, and a brief stop in the town's market square to inspect the walls convinced me that it was the natural stone colour rather than that of a pigmented paint. I was alerted too by the cross-lamination I observed in many of the building blocks – this is a particular type of sedimentary structure imparted during the depositional process. Sure enough the road out of Khenifra then wound through many miles of these distinctive reddish-coloured sedimentary rocks – sandstones with cross-lamination, pebbly sandstones and the still coarser-grained sediments containing pebbles and boulders, called conglomerates.

The sedimentary structures (ripples, sandwaves and chaotica) spoke of powerful rivers, flash floods, debris flows and avalanches. There were very few if any fossil shells preserved, in stark contrast to those found in the shallow marine sandstones used by the Romans in the construction of Volubilis. Instead, there were occasional fossil imprints of branches and tree trunks long since decayed, and smaller black fragments arranged in distinct layers within the sands – former unidentifiable plant fragments that had now turned to coal. These features and the characteristic red colouration that is imparted to the soils of the region as well as to the buildings are recognized the world over as indicative of ancient continental deposits – those of rivers and deserts, and of the alluvial fans and scree slopes that fringe a mountain range under attack from the elements. They are known as the *red bed* association, for obvious reasons, and typically appear in the geological record immediately following a major period of mountain building.

More specifically, red beds of this age, derived from erosion of the Central Pangaean Mountains as well as from other mountain ranges thrown up as the supercontinent sutured together, are known as the New Red Sandstone. A closely analogous but older series of red beds dating from the Devonian Period, scattered across the northern hemisphere lands that once formed the supercontinent of Laurasia, is known as the Old Red Sandstone. The New Red Sandstone ranges in age from early Permian to late Triassic (approximately 280–220 Mya), straddling a period of intense turmoil and change in life on Earth but, because continental sediments are so often monotonously barren of fossil remains, they are not the best place to observe these changes.

What we can learn from these dispersed red beds are two very important aspects of the Pangaean environment. The first is the intense erosion to which the newly formed mountains were immediately subjected, stripping layer upon layer, metre upon metre, and ultimately kilometre upon kilometre, from the uplifted land. Seasonal rain, snow and ice coupled with ferocious winds were the hugely effective agents of erosion. The vast quantities of material removed were deposited by mountain fans and powerful rivers in thick successions of alluvial and fluvial sediment both north and south of the Central Pangaean Mountains. I am continually amazed by the uniformity I have observed in these successions, everywhere from the thick Coconino formation of Arizona, right along the US Atlantic seaboard to the magnificent red cliffs around the Bay of Fundy and the Gulf of St Lawrence in maritime Canada, cutting a swathe across the UK from Devon through central and northern England . . . and then again in parts of Russia, China and South Korea.

The scale of the deposits and the sheer size of some of the boulders held within the conglomerates are clear evidence that very powerful rivers flowed across the continent. Pangaea was enormous and its interior regions undoubtedly dry. Large inland drainage basins with ephemeral lakes would have existed, much like Lake Eyre in South Australia today. These were fed intermittently by torrential rivers – but we have no way of telling whether the deposits represent annual melt water streams or 100-year flash flood events. My geological suspicion favours the latter view.

However, at least some river-borne sediment would have reached the sea. Major drainage to the north of the Central Pangaean Range was directed towards the west, debouching into the main Panthalassic Ocean, whereas that south of the range flowed eastwards into the Tethys Ocean. At this point, you may well ask how on earth we know the flow direction of 250-million-year-old rivers – exactly the question my daughter, Kiah, posed when I was explaining the equivalent red beds near Exmouth on the southern English coast.

The answer lies in the fact that any ripple or dune deposited by unidirectional flow (as in a river) is asymmetric in form, with a gentle slope facing upstream and a steeper avalanche slope facing downstream. So, by careful examination of many such ripples or dunes in a series of red beds, and of the dip direction of the fine lamination they leave behind, the original flow direction can easily be deduced.

The second intriguing environmental fact derived from the New Red Sandstone of Pangaea is directly related to the huge size of the supercontinent and the inevitable consequence that its interior was thousands of miles away from any oceanic influence. At low latitudes in the heart of Pangaea, computer models show that rainfall could not have exceeded an average of 2 mm per year, which effectively translates to absolutely no rain at all for years at a time, followed by rare torrential deluges. Summer temperatures would have frequently exceeded 50°C. This was desert – extreme desert – and almost certainly larger, hotter and drier than the Sahara, Simpson, Gobi or any other of today's great deserts. And what of the evidence in the rocks? Let me explain.

As a young geologist I returned to the UK fresh from a PhD at Dalhousie University in Canada's maritime provinces. It was there that I had first encountered such great thicknesses of New Red Sandstone – deep carmine red-coloured cliffs that gave rise to spectacular rose-tinted sandy beaches. These red beds were undoubtedly fluvial (not desert) in origin. I paid them little heed at the time for my PhD concerned the erosion and reworking of these and other sediments, and their transport into the marine environment offshore Nova Scotia, where they became the deep-sea mud and sand that has remained my research speciality ever since. Back home in the UK, my first job was with the British National Oil Corporation (long since sold off to private interests) as a rig-site geologist. As a newly qualified specialist in deep-ocean sediments I was quietly horrified to find myself on an oil-rig in the southern North Sea drilling into thick Permian-age sandstones, which had certainly not originated in deep water.

The sea all around was cold and grey, icy mid-winter mists hung in the dank air, and the hidden screeches of gulls were caught on the swirling winds. My promised escape by helicopter was not due until the following week. Against this backdrop, it seemed somehow ironic that the very uniform, red-coloured sandstone in our drill cores was undoubtedly of desert origin – I was looking at wind-blown (aeolian) sands. These were the same age as my Nova Scotian red beds but very definitely from a wind-swept, baking hot sand sea in one of the great Pangaean deserts. The clues for this interpretation are in the tiny grains of sand themselves, which become perfectly rounded and polished (rather like frosted glass) due to constant grain-to-grain collisions that occur while they are airborne. In rivers and the sea, such collisions are much muted by the ambient water so that the same degree of polishing does not occur. The sand grains also become size-sorted as they are blown into ripples and dunes, while the avalanche faces of these bed-forms are preserved as cross-lamination, just as in the river deposits I noted near Khenifra.

Larger fragments of rocks or pebbles that lie scattered across the desert are similarly abraded by whirling sand grains under the force of relentless winds. Their windward face becomes somewhat flattened, smoothed and polished until an especially strong gust of wind flips the pebble over and the polishing of a new face begins. Pick up a few of the pebbles from the surface of any desert today and you will soon find a beautifully polished, three-sided example known as a *dreikanter* (Fig. 3). Sure enough, in the pebbly sandstone sections of our North Sea cores we found plenty of Permian-age *dreikanters*, and some that were further tarnished with a coating of black manganese dioxide from long exposure to those past climate extremes.

FIG. 3 Photograph of desert-polished dreikanter pebbles from the Great Pangaean Desert. Width of view 10 cm (Photo by Claire Ashford).

It is interesting that these desert sands – polished, swept and sorted by fierce, hot winds that once blew across Pangaea over 250 million years ago – make almost perfect reservoirs for oil and gas when buried in the subsurface as they are beneath the North Sea. The natural-gas revolution that swept the UK and continental Europe in the late 1960s and 1970s, bringing a new source of heat and power to our homes and factories, was the result of a plethora of gas discoveries made in these desert sands beneath the southern North Sea. Discovered more recently is the giant onshore oilfield located directly below picturesque Poole harbour on England's south coast. This too is mainly reservoired in continental sediments (desert and river sands) of the Pangaean interior.

THE EDGE OF TETHYS

Although there are scant signs of past life from the scorching interior of this vast continent, the extreme conditions having made the preservation as fossils of what life did exist well nigh impossible, some hugely significant finds have been made. These come as fossils preserved in finer-grained sediments deposited in inland lakes and shallow seas, into which ephemeral rivers delivered their mud-charged flood waters. Furthermore, such inland seas were periodically inundated with truly marine waters when earthquakes and earth movements led to significant tectonic readjustments in the supercontinent. Fingers of the Tethys Ocean penetrated far inland. The Zechstein Sea of northern Europe is a prime example.

This sea (Fig. 4) once covered a large area of north-eastern Pangaea, including much of central England, the North Sea – where I had drilled through desert sands, the Netherlands, Belgium and northern Germany. Several episodes of flooding, either through an arm of the Tethys Ocean from the south or via a similar narrow seaway leading from the main Panthalassic Ocean, filled the Zechstein Sea at different times, covering the former desert sand with a vast expanse of relatively shallow and salty water. Along with these waters came a fully marine flora and fauna marching inland towards the continental interior from the edge of the Tethys Ocean. In the organic-rich *Kupferschiefer* (copper shales) of southern Germany, there is a largesse of fantastically preserved fossils that provides a unique window to the late Permian Tethys Ocean. This is one of the few places where we can learn something of the many different types of fish that swam in the Tethys waters of this period – quite unusual species that are now long extinct. So too there are tough-shelled bivalves and brachiopods that could withstand the buffeting of waves along the

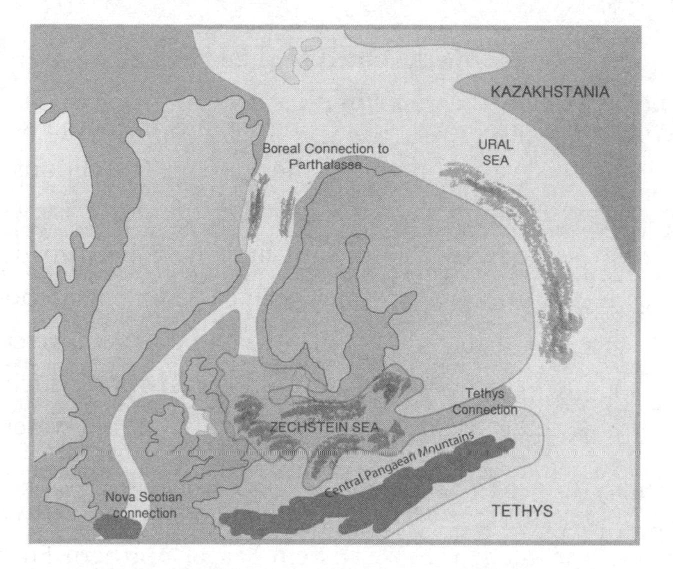

FIG. 4 Reconstruction of the western margin of the Tethys Ocean, where the Zechstein Sea spread across part of what is now NW Europe. Principal areas of evaporite deposition are shown with stipple ornament.

shoreline, and interwoven gardens of delicate fan-shaped bryozoans, porous sponges and multi-armed sea lilies (or crinoids).

But connections with the Tethys were only partial or temporary and, once they were closed off again, the intense desert heat soon returned and evaporation was swift. Hundreds of metres of salt deposits, mainly gypsum (calcium sulphate) and halite (sodium chloride), were left behind as blistering white crystalline flats over the desert sands. The salt mines everywhere from Cheshire in central England to Stassfurt in Germany are a fine testament to this uniquely hostile period of Earth's tortuous history. Once again, any signs of Tethyan life were driven from this inhospitable region.

However, not everywhere was quite so desolate. Lying at a mid latitude south of the equator, there was another inland drainage system that I shall refer to as the Greater Parana Basin. This comprised its namesake in Brazil and the Kalahari Basin in southern

Africa, which were formerly joined together, extending over an area of some 2.5 million square kilometres, roughly comparable in size with the Great Simpson Desert and Lake Eyre Basin, which cover much of eastern Australia today. It was there that three remarkable stages in the history of reptilian evolution have been captured in fossil evidence – a history that was to shape both the Age of Dinosaurs and the Age of Mammals, and hence lead ultimately to our own bit part in this theatre of life.

It was during my most recent visit to Brazil in March 2009, at the invitation of Scottish Development International, that I had the opportunity to venture back in time once more to this period of Earth history. Brazil is a vast country; larger still than was the Great Parana Basin of Pangaea. At first it was a relief to leave behind the high-rise metropolis, the sprawling suburbs and shanty towns that have helped elevate Sao Paulo into South America's number one megacity, as we drove west and south along quiet and well-appointed highways. But after six hours the relief began to pall. We drove through mile after mile of rolling green countryside, hour upon hour of undulating rich farmland, weaving our way along highways and dirt tracks alike, foresters' roads and farm lanes – in search of geology, of course, of any rocks that might have peaked above the landscape and survived the onslaught of tropical weathering that yields such deeply rich and reddened soils. Our search was a long one. The whole of the State of Sao Paulo, it seems, is covered with sugar cane plantations – an area twice the size of Britain – unmoving fields, still in the heavy humidity of the day.

These proud plantations of such insubstantial plants have fuelled a bio-revolution in Brazil – a nation of 200 million people and almost as many vehicles. But all those made today run on flex-fuels, either gasohol or pure alcohol, sold side by side at the pumps. Our minibus was wholly alcohol-fuelled, so that as we struggled up the steeper hills, the exhaust fumes that inevitably filled the air were sweetly alcohol-scented rather than heavy with diesel oil.

We drove for several more hours, still through these relentless rolling hectares of sugar cane, before a subtly more elevated relief brought respite. Now there were cattle prairies and plantations of eucalyptus forest, sluggish streams and wider rivers, brown and laden with sediment, that support wilder untamed vegetation, a profusion of growth and fertility, the air laden with sweet, almost sickly blends of scents from the colourful blossoms that abound. Above the glorious red hibiscus, darting hummingbirds, and the noisy buzz of so many insects, soared buzzards and wedge-tailed eagles and, higher still, on warm gyres, the vultures circled.

At times I wondered at our chances of actually finding any rocks outcropping beneath this impressive sea of green, but I should not have worried. Professor Dimas Dias-Brito, my geological companion and guide from Sao Paulo State University at Rio Claro, knew well the treasures of the Great Parana Basin and in just which quarry or along which scarp they were to be found. At last we found a hidden valley of ancient desert sands, now oozing with black asphalt, the giant cross-bedding of ancient wind-blown dunes picked out so clearly by an alternation of black and white lamination. Even the escape burrows left by fossil insects are still visible, and the collapse structures of sand-flow down the face of over-steepened dunes. Once quarried for its heavy oil, the place is now quite abandoned, so that the rich vegetation of today's world has once more taken hold, the boggy valley floor steaming with biting insects and hidden snakes, and the rocky sides pitted with myriad holes, home to gigantic brown Brazilian hornets – of which Dimas and his colleagues were distinctly wary.

So, indeed, we had arrived at the desert edge – the Parana Desert I shall call it – of this former great basin. From here the story unfolded.

First we found fossil archosaurs. These are the root stock from which evolved the myriad swimming reptiles that came to dominate the Tethys seas, as well as all the many dinosaurs who patrolled the shores and ruled supreme across the breadth of continents for countless millennia. Among the ever-growing number of dinosaur species that paleontologists seem to report almost annually, from those the size of bantam chickens to the fearsome *Tyrannosaurus rex* and lumbering sauropod giants, there was the one that later evolved into the first-ever bird species. Before any of this could occur, primitive archosaurs had to survive the worst calamity of all time – the great Permian extinction event, which is the subject of my next chapter.

Even more important, at least from an anthropogenic perspective, is the second fossil treasure to be found throughout the Great Parana Basin – the cynodonts. These represent a different reptilian lineage from the dinosaurs. They were synapsids rather than diapsid reptiles (the dinosaurs). This classification is based on the number and arrangement of holes (apses) behind the eye socket in the skull. Synapsids have a single hole low down on either side of the skull, whereas diapsids have two. Although this may seem a rather minor and almost bizarre differentiation, it was actually of enormous significance. The diapsids evolved into dinosaurs and marine reptiles and so came to dominate both land and sea for many years to come. One group later went on to evolve into birds. Synapsids may have played second fiddle for a while but one branch was later to evolve into mammals.

Cynodonts were the last of the synapsids to evolve, more or less as our story of the Tethys begins, and one of the most successful. They began as wolf-like carnivores, which adapted easily, spread rapidly to all kinds of habitats around the world, survived the impending Permian extinction event and lived on for another 70 million years.

But before they finally died out they left descendants that evolved into mammals and eventually, of course, into humans. In fact, the first discovery of a cynodont (or at least its published description) came not from Brazil but from the other end of the Great Parana Basin, in Luangwa Valley, Zambia. Two other colleagues of mine (Steve and Anna Tolan) run what must be one of the most remote nature reserves in the very heart of Africa, high up the Luangwa River, a tributary of the Zambezi – five days of arduous driving by Land Rover from the capital, Lusaka. They know that they are sitting on a veritable gold mine in the early evolution of our direct descendants, perhaps even the missing link between cynodonts and true mammals. But there have been no serious paleontological studies, or even cursory visits to the region, since an expedition from Oxford University led by Dr Tom Kemp some 40 years ago.

Exactly the same *Luangwa* species of cynodont was discovered a few years ago in Brazil. Perhaps there is another 'gold mine' to uncover beneath these sugar cane plantations... which I must leave for another time, another longer expedition perhaps.

Before leaving Brazil, our final visit was an excursion to the Irati Formation at Partecal. This was quite spectacular – a huge quarried region, first cutting through a mudrock that produces more than half of Brazil's ceramic clay, and then into an amazing tiger-striped alternation of dark and light-coloured rocks. As we worked our way through these Permian 'rhythmites', as they are called locally, we slowly uncovered more and more of a unique story. The sediments were deposited in part of a huge inland sea, in the very hot and arid deep interior of the Pangaean supercontinent around 250 million years ago. This was another example, similar to the Zechstein Sea but much larger in scale, where marine waters had flooded into the very heart of Pangaea. The Tethys Ocean, or perhaps a long arm from Panthalassa, had fingered its way inland

and brought with it an array of distinctive marine plants and animals.

We found rocks covered with thousands of tiny ostracod fossils. The low species diversity but very high numbers was clear evidence of unusual salinity – either extremely salty or brackish to fresh waters. These were interlayered with black organic-rich shales, indicative of oxygen-starved waters. And then, to huge excitement, we began to find a very different fossil type, the scattered remains of ribs and vertebrae – some still articulated and almost completely intact – of mesosaurs (Fig. 5). This was our third gem of reptilian evolution, representing the first aquatic reptiles to evolve from their wholly land-based ancestors. They resembled small alligators, which, according to some paleontologists, fed on plentiful blooms of plankton in the surface waters using their teeth as primitive baleen filters. Other paleontologists have suggested that their principal diet was one of small fish and crustaceans – ostracods, perhaps. There is still debate as to whether they were actually marine or lived in large inland seas. Although the mesosaurs may have paved the way for a return to water, the group itself did not survive the cataclysm that was soon to follow – or did they?

Because relatively little is known about this important group, my colleague from Heriot Watt University, Patrick Corbett (who was with me in Brazil on this excursion), announced his intention to undertake a second PhD on the subject of mesosaurs when he retires from his current professorial role.

Of course, our intended one-hour stop ran into several hours and the day marched on. And all this time we were being watched closely by a small and inquisitive burrowing owl who lived somewhere in the upper quarry face, where the deeply weathered rock became soft enough to excavate. Much later, we watched the sun set from a high vantage point on the edge of that ancestral basin,

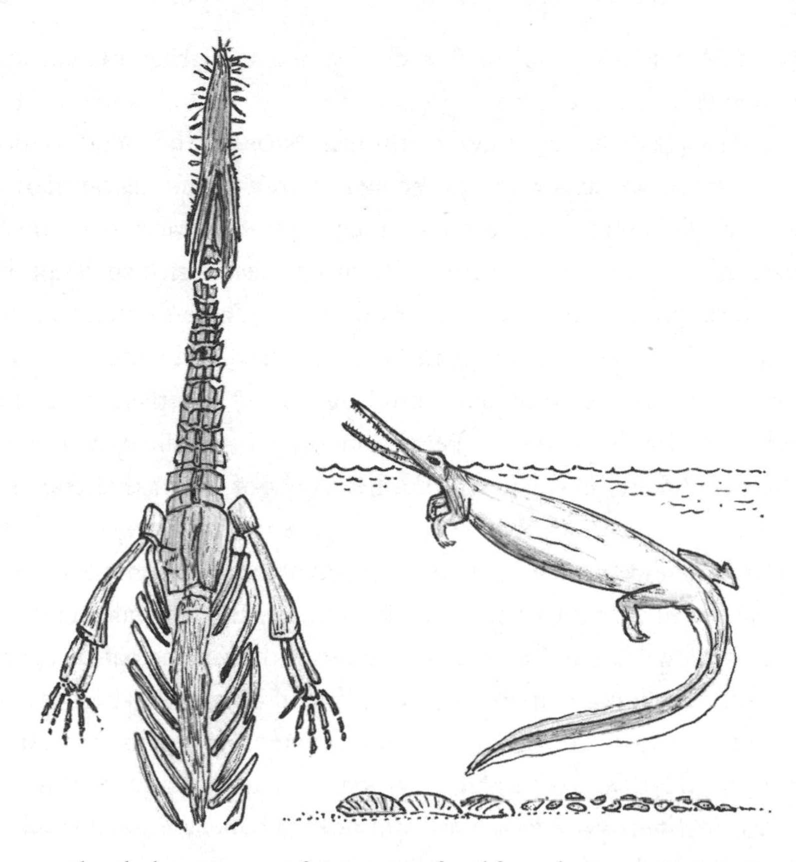

FIG. 5 Sketch showing part of a Mesosaur fossil from the Parana Basin in Brazil. Reconstruction of Mesosaur swimming (length approximately 0.6 metres).

while our memories and excitement of the day mellowed with cold beer, conversation and camaraderie.

HOT SPOTS AND RIFTING

This chapter has sought to address something of the assembly of the Pangaean supercontinent, the nature of its interior and the waters that lapped at its shores. The Tethys Ocean lay to the east while all its other margins abutted on Panthalassa. It was only when Pangaea had fully assembled around 260 million years ago that the Tethys became fully demarcated as an ocean in its own right, with an arc of scattered islands separating it from Panthalassa still further to the east. Periodically, as a result of tectonic activity (continental plates rising or falling) and/or due to changes of sea level, the Tethys extended long arms into the interior of Pangaea such that parts of the continent were flooded with shallow seas. These marine waters brought life from the Tethys to an otherwise desolate interior. But, as often as not, these ephemeral seas soon dried up, the seawaters evaporated and left behind their dissolved salts (sodium chloride, calcium sulphate, and others) as thick deposits of what are known as *evaporites*.

It is apparent from all that we now know about the interior heat engine of our planet that the ultimate foundering of this supercontinent was inevitable. Pangaea probably only remained intact for a few tens of millions of years at most before it began slowly to break up and then to drift apart. In one of the most remarkable and least understood episodes of Earth history, deep fractures or rifts appeared almost simultaneously throughout the supercontinent (Fig. 6). Basaltic lavas poured forth in great profusion, the relics of which can be found today, especially around the margins of the Atlantic, as ancient flows and intrusive dikes.

Lying beneath the Earth's hard and mostly rigid exterior crust, the inner mantle and core are far from passive players in affairs at the surface. Natural heat contained within the core is the principal heat engine driving convection in the mantle and ultimately the movement of tectonic plates. Rising plumes of hot mantle are known to occur as isolated hotspots beneath the crust, such as those beneath Iceland and Hawaii today, spewing out lavas and volcanic ash from which whole islands and island chains have been built as oceanic plates move over them. The Azores, Cape Verde, Mauritius and the

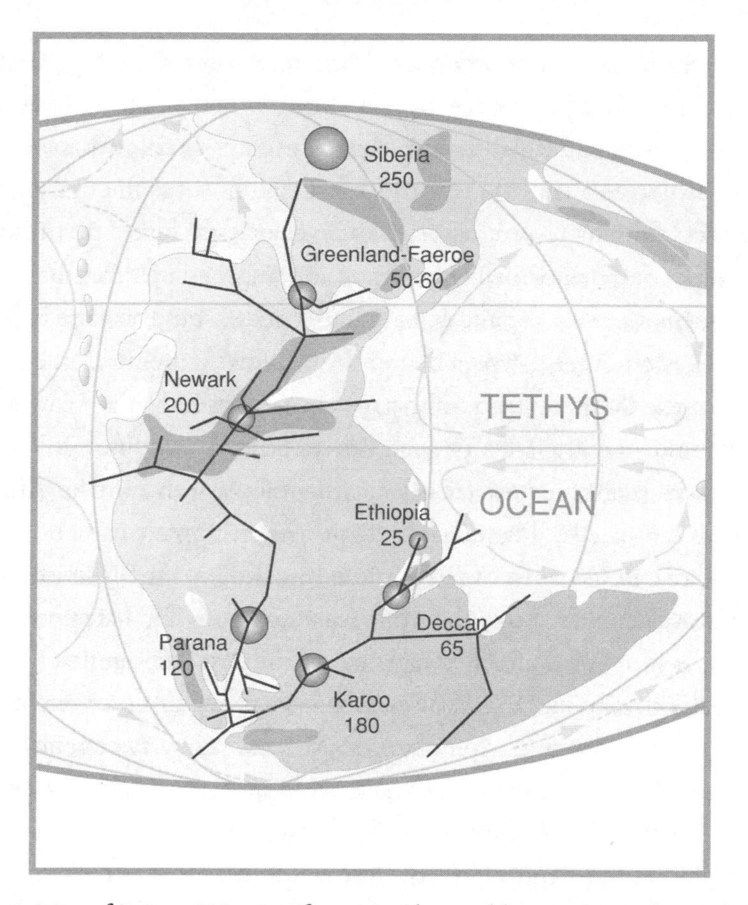

FIG. 6 Map of Permo-Triassic rifts across the world at 240 Mya, showing the intense fracturing of Pangaea that preceded its eventual break-up. Circles depict major igneous provinces that gave rise to extensive volcanic activity, with approximate dates in Mya. Younger igneous provinces are associated with more recent rift episodes.

Galapagos islands are other well-known hotspots. But all of these occur beneath the oceans. What happens when a hotspot forms beneath a continent?

Inevitably, a supercontinent the size of Pangaea would have drifted over one or several hotspots. Indeed, it is now thought that there may be a causal relationship that induces hotspots to develop beneath such large continental plates. In either scenario, the excess heat generated from a single hotspot causes the earth above to bulge upwards until a radial fracture pattern develops, typically with three principal fractures known as a triple junction. Some or all of these lines of crustal weakness founder to produce a rift valley. Lavas are forced up through the cracks and help widen and deepen the valley still further. It is into rifts of this sort that the ancestral Tethys flooded, but this time at least some of them were permanent and deep, allowing new oceanic crust to form the floors of deep but narrow ocean basins.

The nature and effects of such hotspot activity at present and in the more recent past are clearly seen in the eastern horn of Africa. The Abyssinian Highlands centred on the Ethiopian capital Addis Ababa, rise in places to over 4000 metres, and are the result of long-lived domal uplift above the Ethiopian hotspot. The three arms of the triple junction so formed have led to the East African Rift Valley, the Gulf of Aden and the Red Sea. The last two of these are incipient oceans in the making, while the East African Valley system also has the potential of opening up as a new ocean and splitting Africa apart. I shall return to this point in the final chapter.

So it was for Pangaea. Rifts were beginning to form as mighty cracks in the very fabric of the supercontinent. Both the Tethys and Panthalassa were poised to flood inwards if and when the rifts deepened and opened further. But, for all these mighty forces at work, the Earth changes with an almost imperceptible slowness and extreme lassitude. For the moment, therefore, we must be content to pass into the next chapter with a world deeply shaken and fractured but not yet torn asunder.

3

Extinction, Evolution and the Great Cycles of Life

time runs
like a broken river
dragging the heavy dead,
trees uprooted
from their whisper, everything
is racing toward hardness:
dust and autumn,
brooks and trees,
water will pass away:
then a mineral sun will gleam
over all these stones...

From *Stones of the Sky*
by Pablo Neruda
(translated by James Nolan)

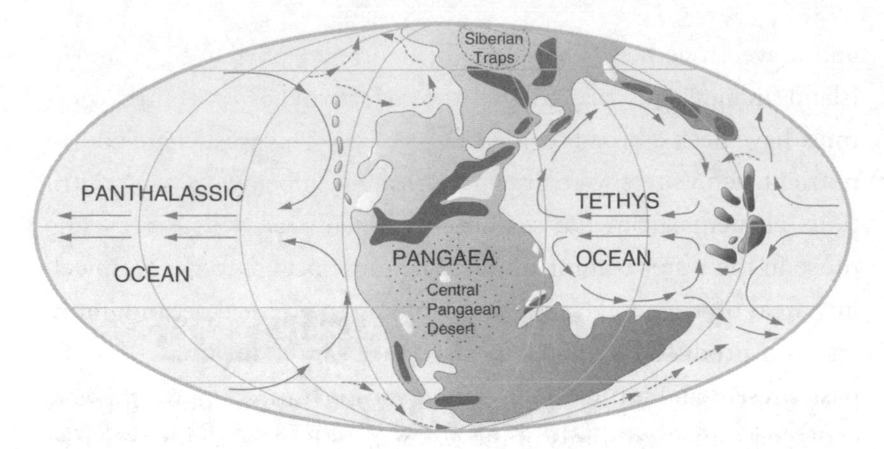

Early Triassic Tethys map (240 Mya). Global reconstruction with oceanic circulation, also showing principal mountains, deserts and evaporite deposits as before (Ch 2). Note location and extent of Siberian traps – this extensive vocanic province was active at the PT boundary 250 Mya.

Impending disaster. What disaster? What portending changes were afoot at the dawn of this new ocean world? All seemed well in the early Tethys, at least to the casual observer. The ocean commanded a prime equatorial position, sparkling under a benign climate and brimming with life. It was then many years beyond the ice-house conditions that had gripped much of Gondwana before its ultimate fusion into the supercontinent of Pangaea. Tethys was a broad expansive ocean with surface currents that spun in two great gyres, clockwise to the north of the equator, anticlockwise to the south. This pattern is the inevitable result of a force induced by the natural spin of the Earth about its own axis – known as the Coriolis force. The same force that causes aeroplanes to deviate from a straight trajectory as they fly, and the currents of today's oceans to spin in just the same direction as they always have. Or, as some would have it, the bathwater to spin down the plughole in one particular direction!

There would have been a strong westward-directed equatorial current as the northern and southern arms of the supergyres combined and flowed together. Where this dual current pulled away from the island archipelagos that fringed the outer reaches of Tethys, there must have been a broad region of upwelling. This was where cooler, nutrient-rich waters were drawn up from deep below, stirring nutritious elements from the sea-floor sediments as well as recycling those in the water column above. The same occurs today in upwelling areas off Peru and Namibia. According to physical oceanographers who produce computer models of ocean circulation – for the past, present and future time – this equatorial upwelling would have extended some two-thirds of the way across Tethys. There were two further regions of upwelling, in the extreme north and south of the ocean respectively.

It was in these regions, in particular, that life abounded. The sunlit waters at the ocean surface were teeming with plankton, which formed the base of a complex food web in a hungry ocean and therefore the primary support for most of life in the late Permian Tethys. Seed-shrimp ostracods were among the army of tiny primary consumers harvesting an even tinier phytoplankton, or scavenging the organic detritus that rained to the sea-floor. Here they were joined by trilobites, making their final cameo appearance in a very long and distinguished march through the entire history of Palaeozoic seas, which endured for more than quarter of a billion years. Trilobites were another type of arthropod, a whole subphylum in fact, that looked rather like marine wood-lice, but with some species growing to over half a metre in length. Alongside them, minnow-sized, eel-like conodonts swam or slithered in their millions. Most species were quite unlike those of today – they were all of the Palaeozoic ('ancient life') Era. Even the rich reef communities that flourished along the shoreline of volcanic islands were unique, almost alien, with giant lamp shells, sea-lily crinoids and countless schools of strangely different fishes from a bygone era darting in and out of rocky crevices for food and safety. The top predators were sharks. So successful was this particular evolutionary masterclass that they have truly withstood the test of time and still rule supreme in many ocean waters today.

However, it was not long after this strange but vibrant scene at the heart of the ocean, not long after the birth of the Tethys, that the world was to change forever. That was 250 million years ago – just 6 million years that Palaeozoic life was allowed to flourish in a new ocean before the bells of time chimed for change. Imagine a catastrophe so severe that life on Earth was but a whisker away from complete annihilation. That was exactly what happened. It was undoubtedly the greatest disaster of all time – at least for which we have good geological evidence. In an extinction event of gargantuan proportions, an estimated 90–96% of all species on Earth vanished, the loss at sea being even harsher than that on land. Though scientists are still sifting through the tragic remains of this cataclysm for evidence of its cause, we do know that the oceans took nearly 5 million years to establish their recovery on a secure footing. A completely new-look Tethys would eventually emerge and life would flourish as never before. But for now it was hanging by the merest thread.

Evolution and extinction have always been closely intertwined in the web of existence. The passage of species through geological time – their appearance, development and demise – weaves an extraordinarily complex pattern that is not easily understood. At certain times and places in the past, conditions have favoured rapid adaptation and change, great radiations of new creatures such as occurred in the Cambrian explosion of life, long before the Tethys Ocean was born. Other such periods of change are reflected at the beginning of each new geological era, for it is largely on this basis that the eras and periods have been defined. Other times have witnessed the rapid and simultaneous demise of large numbers of species, and even whole families of related genera. These are the mass extinction events of which there have been several in Earth history (Fig. 7). They include the great Permian extinction (Permian – Triassic or PT) event that we are about to discuss, as well as the KT event (Cretaceous – Tertiary) that brought an end to the dinosaur reign, and which has so engaged both the public and scientific community alike. But such fascination inevitably leads to wild speculation and simplistic theories. The truth is undoubtedly more complex.

HOW MANY, HOW FAST?

In considering the nature of past extinctions, an obvious question to ask is 'How many species of plants, animals and other organisms have ever existed on the Earth?' Furthermore, how long do individual species generally survive the challenge and competition presented by the environment in which they live? The first step in considering these questions is to determine how many species are actually living today – but, in fact, there is no easy answer even to this. A generally accepted estimate is somewhere between 5 and 10 million. Of these we can be confident in naming many of the larger animals, about 4500 species of mammal for example, but only guess at around 1–2 million types of insect.

There is so much we have yet to learn about life on Earth, so many uncharted territories – insects of the tropical rainforest, bacteria that live in the sediment beneath the sea-floor, weird and wonderful creatures living near steaming hot springs on the sea-floor (black smokers), or lost in the silent darkness of deep ocean space. Every year, at least 10,000 new species are named for the first time – and an estimated 30,000 become extinct. This currently high rate of extinction we know to be largely due to human activities.

Estimating the numbers of species in the past is an even more intractable problem, especially when palaeontologists calculate that more than 95% of past life was never fossilized in the first place. This is because the odds are firmly stacked against any organism ever making it into the fossil record – the microbial army of decomposers and waste-recyclers has always been highly proficient. And where the microbes perchance fail, then there are chemical and physical extremes of burial and lithification (turning to rock) still to endure. Even that which is fossilized we believe to be quite unrepresentative – shellfish (molluscs and brachiopods) are relatively common as fossils because their tough shells are more easily preserved, whereas insects are rare (except within amber – a fossilized tree resin), and wholly soft-bodied organisms (jellyfishes or worms, for example) rarer still. An admittedly speculative estimate for the total number of species ever is around 750 million. Palaeontologists have rather greater success in counting the numbers of larger groupings, such as families or orders. From this information it is possible to chart the rise in numbers of families through time – to recognize periods of rapid increase and decrease, as well as different evolutionary groupings that characterize the principal geological eras.

Clearly, evolution progresses at very different rates. Rapid change and speciation occurs when a virgin area is colonized, such as the Galapagos Islands – isolated volcanic edifices that pushed above the ocean surface 1–2 million years ago as the direct result of localized hotspot activity. Soon after plant life had taken hold, closely followed by insects, the first finch-like birds flew over from South America and rapidly took advantage of this newfound bounty. They quickly evolved into 14 different species, some to capture small insects, others large ones, some to crack nuts, others to forage for seeds, and so on. At the other extreme, some species seem locked in time, living fossils from a past world. The shy *Coelacanth*, now rarely found in the depth and darkness of the Indian Ocean, is remarkably similar to some of the earliest true fishes. The *Tuatara* lizard of New Zealand has barely changed from those that scampered across rocks by the Tethys shoreline some 225 million years ago. Perhaps most remarkable in this way is the *Lingula* brachiopod, a type of shellfish that burrows into the soft sea-floor for safety, which time and evolution have forgotten completely. I have found exactly the same genus fossilized in Ordovician mudstones from South Wales, as can be found burrowing into the mud of shallow sea-floors today, more than 500 million years later.

Within this enormous variation, it turns out that most species have lasted for fewer than 10 million years. Mammals, fishes and beetles are short-lived species, surviving for only 1–3 million years on average, and some insects may last for as little as a few thousand years. Shellfish, on the other hand, are longer lived, many lasting for 10–15 million years. It is often the simplest forms, microscopic plants and algae floating on the ocean surface, that last for longer, showing little change over 20–30 million years. There appears to be a natural survival duration that is characteristic for individual species or groups, a biological clock silently ticking away the millennia of existence. This is an important perspective with which to view species extinction through time.

SKELETONS FOR SURVIVAL

Prior to the great radiation of life in Cambrian times, around 550 million years ago, little is known or preserved. Primitive single-celled and multicellular organisms of the Precambrian (Proterozoic and Archaean Eras), from which we all ultimately evolved, must have undergone their own dramas. Change and upheaval in the oceans and atmosphere, volcanic outpourings and bombardment from space, might many times have caused almost complete extinction of life on Earth. Perhaps there was even a closer call than that which occurred at the PT boundary. We shall probably never know, but somehow life clung on, eventually with the remarkable success we see today.

Evolution was even slower than Charles Darwin had originally believed, and was played out entirely within the seas for around 4 billion years. There must have evolved a hidden and well-developed life waiting to burst onto the scene, as we know from rare finds of multicellular soft-bodied fossils such as at Ediacara in South Australia. The development of animals with skeletons, which leave a much better record of their existence, happened around 545 million years ago and marks the beginning of the Phanerozoic ('visible life') Era. Each subsequent period through the whole of this Era is marked by a new stage of evolution at the outset and culminates in a more or less catastrophic extinction event (Fig. 7). However, we can detect no single major environmental change that took place across the Proterozoic–Phanerozoic boundary and which may have precipitated the dramatic evolutionary event that was to follow. There was a mild equable climate, which contrasted with at least two earlier glacial episodes: a global warming and rising sea-level providing more shallow sea ecological niches that perhaps contributed to the great radiation of life at this time. Probably more likely as causative effects were chemical changes in the composition of seawater (and/or the atmosphere) at the time. But all this is quite obscure and presently the subject of further work.

Whatever its ultimate cause, a diversified and relatively advanced flora and fauna rapidly colonized the Cambrian seas. This event is referred to as the 'Cambrian explosion', and heralded the appearance of calcified algae, hard-shelled bivalves, gastropods and echinoderms, together with the first corals and sea lilies. Long-extinct but highly

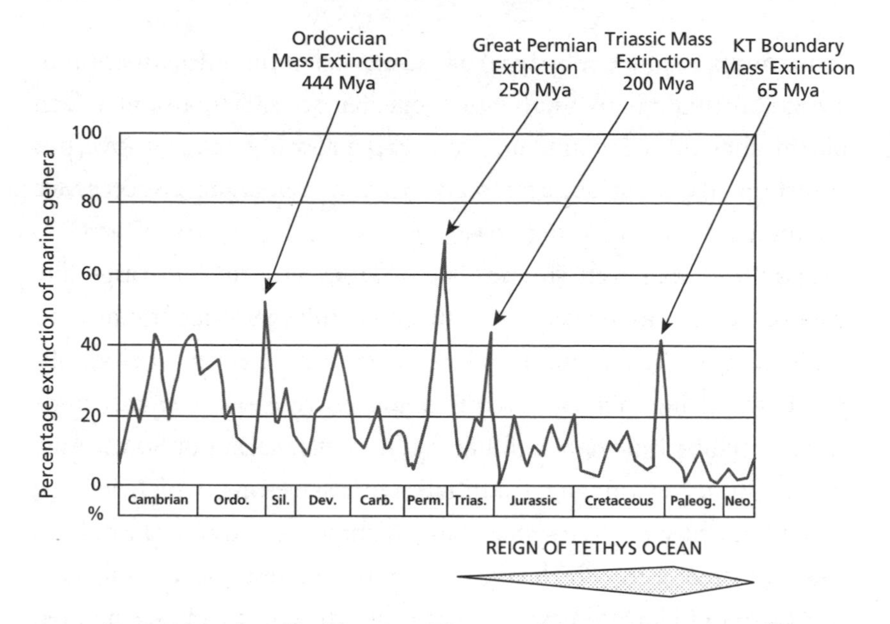

FIG. 7 Extinction chart for Phanerozoic time, showing percentage of genera that became extinct at different times over the past 550 million years of Earth history. Note at least four past Mass Extinction Events.

successful groups such as trilobites and the small, colonial animals called graptolites swam, crawled or floated throughout the marine world. Species came and went with a hitherto unknown geological rapidity – ever more complex and exotic, each trying to exploit a new and unused ecological niche.

Coupled with this profuse radiation of species, and equally abruptly, these new organisms evolved hard parts – variously made of lime, opaline silica, or chiton (a hard organic molecule). Bone material of phosphate, as modern bone is, had yet to be developed. Alongside the evident question of why animals appeared so suddenly at the beginning of the Cambrian period, is the similarly intriguing one of why they first acquired skeletons – hard external skeletons (or shells) that is. In evolutionary theory we must seek to understand what selective advantage this provided. Indeed, there are a variety of possible and partial explanations: it provides protection against fierce ultraviolet radiation, especially near the ocean surface and in shallow waters; it prevents rapid drying out in intertidal environments; it provides protection from predators; and it gives a framework to support growth in size and the attachment of muscles. At this point, we must be content simply to observe the occurrence of hard parts and be delighted with the added bonus that mineralized skeletons are 1000 times more easily preserved as fossils than is soft tissue. Without this relatively easy preservation of fossils, there would be little to tell of life in the Tethyan Ocean, which was yet to appear towards the end of the Permian period.

The diligent work of palaeontologists has ensured that the evolution and demise of life through Phanerozoic time is much better known than before. It can be neatly divided almost in two by the great Permian extinction, which occurred at the end of the Permian Period, 250 million years ago. Four other major mass extinction events, as well as about ten smaller ones, are clearly recorded in the fossil record. Arguably, we are in the midst of another mass extinction, mainly of our own making. There is almost an uncanny cyclicity in the timing of these events, at least since the great Permian extinction, with one occurring every 26–30 million years. But the spacing of cycles is not perfect and the potential causes hotly disputed.

Inevitably, perhaps, there has been much popular myth surrounding mass extinctions, often fuelled by a quasi-scientific desire to find an easily understood explanation. For none is this more evident than for the KT extinction event. Much of what has been written is really quite fanciful and unscientific, as I shall argue in a later chapter. In reality, the explanation for the PT, KT or any other extinction event, is much more likely to involve a complex combination of natural biological factors, such as disease, competition, over-population and extinction rates, and external environmental drivers, including climate change, sea-level variation, ocean – atmosphere composition and extraterrestrial events.

FINAL FLOURISH

But let me return to the late Permian period at the end of the Palaeozoic Era. Before the mass extinction that marks the end of the Palaeozoic, life was flourishing as never before across the whole spectrum of ecological niches. Certainly there were harsh environments and extreme conditions, as in the burning-hot interior of Pangaea or on the dazzling white salt flats at the edge of the sea, just as there was healthy competition for survival; but that life flourished is not in doubt. In order to more fully capture evidence of the picture of vibrancy and variety that existed just before the end, I have travelled to three very different parts of the world – the Ural Mountains in Russia, the Northwest Territories in Australia, and finally the Guadalupe Mountains of Texas and New Mexico.

For me the Urals are a strange and surreal mountain range. They stretch in an almost unnatural straight line for over 2000 kilometres across the plains of western Russia, fading out before reaching the Arctic Ocean in the north and similarly disappearing beneath the lowlands of Kazakhstan in the south. They formed as the final constructional phase of the Pangaean supercontinent, when the eastern parts of Russia and Asia collided with Europe. Steadily eroded over the ensuing 260 million years, they now express a much subdued relief with nowhere exceeding 2000 metres in height. Whenever I have flown across this range I am reminded of some gigantic earthen dam holding back the vast and inhospitable Siberian wastelands from rolling hills and fertile farmlands to the west.

About halfway along the Urals on their western flank lies the city of Perm, a Soviet industrial powerhouse of some 1 million inhabitants that sprawls along both banks of the Kama. In almost all respects it is quite unlike the city with which it is twinned in the UK – Oxford! However, there is a certain charm that derives from its broad, magnificent river flowing south into the Volga and forming part of an amazing and fully navigable network of waterways joining the Black, Caspian, Asov, White and Baltic seas. It was in the countryside surrounding Perm that Sir Roderick Murchison, a pioneering 19th-century British geologist, working alongside the Latvian scientist Count Alexander von Keyserling and Edouard de Verneuil, a French palaeontologist, first encountered a completely new rock series of the same age as the Zechstein evaporite series in Europe, and mentioned in the preceding chapter. These sediments, however, were fully marine, richly packed with all kinds of fossils, and were used by Murchison to erect a new division of geological time, which he called the *Permian Period.* Highly contentious at the time, it did not gain international recognition as a formal geological system in its own right until 1948.

During the late Permian Period, the region of Perm lay under shallow marine waters beneath a northern arm of the Tethys and therefore yields important information on ocean life during that time (Fig. 8). Trilobites were still the dominant scavengers on the sea-floor as they had been now for nearly 300 million years, joined by marine scorpions and shrimps that swam overhead. Brachiopods (or lamp shells), a sort of prehistoric variation on the many different types of shellfish we eat today, were another dominant group found in the Perm strata. They were mainly sessile forms (attached to the sea-floor) with tough shells of calcite, coming in all shapes and sizes, boasting smooth, ribbed or spiny exteriors, and often living in large numbers cemented together on rocky substrates. They lived alongside bivalve molluscs, gastropods (sea snails), crinoids and branching networks of bryozoans (sea mats and sea nets). Distinctive members of these reef-like colonies were great clusters of cigar-shaped fusulinids, often numbering many hundreds of individuals. These are now extinct members of single-celled foraminiferans, a hugely successful order of protists that count among the giants of the single-celled world. Growing up to 2 cm in length, fusulinids look something akin to babies' fingers.

Seeing the fusulinids in Perm brought a smile to my face. Over 35 years ago, as an undergraduate at Cambridge University, I mounted a geological expedition (or summer holiday!) to the Greek island of Hydra. We discovered several important new localities with abundant Permian fusulinids on the then remote south-eastern coast, together with corals, brachiopods, molluscs, crinoids and bryozoans. This early brush with the Tethys Ocean led to my very first scientific publication in 1975, although I have barely looked at a fusulinid since. Hydra boasted a more subtropical fauna than Perm, similar to that of the broad equatorial shelf seas that covered much of what has now become the Middle East.

Along a different and southern shoreline of the Tethys Ocean, these broad shelf seas flooded across parts of Pakistan and India, which have since been caught up in the mighty plate collisions that created the Himalayas, and even onto the northern shores of Australia. Here was a more temperate and generally wetter climate than for most of Pangaea, as all around the Tethyan rim there would have been a greater chance of coastal rainfall than in the dry interior. Where large rivers entered the sea, swamp-like conditions spread across the delta and coastal plain, much as we see today. I have seen evidence for these Permian coals in many parts of the world, but they are especially impressive in north-west Australia. The same age coals and associated plant fossils occur in other parts of Australia (New South Wales, Victoria, Tasmania) as well as through India, South Africa and South America. Interestingly, the plants of that time were unique and most would be quite unrecognizable in our world.

Standing amidst this late Permian fossil forest of north-west Australia is a quite magical experience. I can almost hear the hum of insects lost in time; and looking out over the broad blue Pacific before me, it is not difficult to imagine the vast Tethys Ocean instead. This was the beginning of the end of the Age of Forests, a period of time that lasted for over 100 million years from long before the story of the Tethys begins. It was a time of huge importance in evolutionary trends and of a significance yet to be fully unravelled for the impending catastrophic extinction that was about to descend. Furthermore, the great forests of the Carboniferous in Laurasia and Permian in Gondwana (southern Pangaea) have since yielded the greatest abundance of coal deposits throughout the world, unwittingly fuelling our own industrial revolution in the 18th and 19th centuries and contributing the most of any natural resource to human-induced climate warming today.

These magnificent forests made a clear and unequivocal statement – the singular dominance of the marine realm as home to life on Earth was well and truly over. For as many as 200–300 million years before the time in which I was standing (late Permian), bacteria, algae, lichens and fungi had been courting with increased exposure to air and had already begun to decorate the once barren land. Seaweeds developed a tough envelope that allowed some rigidity without the support of water as they crept out along the shoreline fringe of an otherwise desolate landscape. Before long the continents were as green as the seas were blue. The advent of this colonization by land plants was to have a major and irreversible effect on the whole global environment.

This brave new world of plant life on land meant that much of the carbon dioxide in the atmosphere was rapidly converted to

FIG. 8 Late Permian marine diorama based on Tethyan fossils: 1 Nautiloid, 2 Crinoids (sea lilies), 3 Rugose coral, 4 Spiny brachiopod, 5 Trilobite, 6 Bony fish, 7 Bryozoan, 8 Blastoid, 9 Glass sponges, 10 Brachiopods (lamp shells), 11 Stromatolite.

oxygen as a by-product of photosynthesis. Eventually a more or less steady-state oxygen level was reached at which animals with lungs could live on land. Through a symbiotic relationship with bacteria, plants acquired the ability to fix nitrogen directly from the air. Another innovation, the unique system of evapo-transpiration in vascular plants, by which water is transported from the soil to the very top of the plant and thence into the atmosphere, greatly influenced patterns of rainfall, global temperature and atmospheric circulation. Soil itself only came into existence following the accumulation of leaf litter and other plant debris on a substrate broken down by increased levels of mechanical and chemical weathering. Nutrient cycles, which had once been firmly rooted in the oceans, had changed forever.

Once the bridgehead had been established by plants, so pioneers of the animal world could venture forth. By the end of Permian time the great forests, such as these in north-western Australia, were alive with a veritable army of insects, swamp-loving amphibians and early reptiles. It is here that we find the distant ancestors of millipedes, centipedes, scorpions and spiders. It was in these early forests that insects first took to the air and so rapidly evolved into the realm of nightmarish giants, especially during the preceding Carboniferous Period. Fossil mayflies and dragonflies have been found whose wingspans reached an amazing 70 cm. It is interesting to note that such magnificence has never since recurred in the insect world – a fact that is most likely related to the extremely high oxygen levels reached in the Carboniferous and Permian.

It was the evolution of plants that made this world possible. Consider a moderate-sized tree today with an ageing trunk covering a ground area of 6 square metres. It has been estimated that the total 'surface area' made available by the mature tree is around 11,000 square metres, an 1800-fold increase in ecospace available for other organisms to colonize – including the tops and bottoms of leaves, twigs, bark, the underside of bark on dead branches, and so on. The Permian forests boasted many such trees and other smaller plants, the giant lycopods, horsetails and ferns, spore-bearing varieties so well known from the Carboniferous period that came before, as well as newly evolved seed ferns such as *Medullosa* and *Glossopteris*. The fossilized remains of this latter genus, in particular, are common throughout the widely separated southern hemisphere continents today – South America, South Africa, Australia and even Antarctica.

This, as mentioned earlier, was one of the key fossils, together with the *Mesosaur* reptiles I came across in Brazil, that Alfred Wegener used in order to show that these disparate landmasses had formerly been joined together and so to support his theory of continental drift.

For my final glimpse of life before the end, I made a short visit to West Texas and New Mexico and stood atop the gleaming white crags of the Guadalupe Mountains as they slowly turned to rose pink in the evening sun. The view across Delaware Basin and Midland Basin beyond is quietly breathtaking in its scale and tranquillity, but this is as nothing to the awesome tale within the rocks. For these spectacular sequences of reef limestone are as if fossilized in place from 250 million years before. In the Permian, they rose 600 metres from the sea-floor to the water's surface, fringing shallow seas that had invaded Pangaea along its western margin, just as now they rise from the exhumed valley plains to the sky.

El Capitan Reef, as it is known locally, extends for some 750 km around Delaware Basin. More than 350 species of all different kinds of organisms have so far been identified from the fossils preserved in the reef itself, as well as in the forereef talus and backreef lagoonal limestones. Colonial reef-forming corals, so familiar in today's tropical waters, had not yet evolved, so that this whole gigantic framework was composed of colonies of bryozoans, calcareous sponges, spiny brachiopods, and green algae, together with gardens of stalked sea lilies and blastozoans (Fig. 9). Just as the forests on land provided ecospace for others to use, so it was with these oases of the Permian seas – myriad fishes, shrimps, jellyfishes, coiled ammonites and nautiloids grazed or captured prey with their own special adaptations, while sharks patrolled at the summit of the food chain.

FIG. 9 Photographic detail of marine crinoid (sea-lily) fossils typical of the late Permian Tethys Ocean – width of view 10 cm (Photo by Barry Marsh).

THE END OF AN ERA

There is little doubt that something quite extraordinary took place around 250 million years ago. An estimated 50% of all known families disappeared forever, which translates to an astonishing 90–96% of all species on Earth – most of which were still ocean-bound at this time, so that extinction was most keenly felt in the marine realm. Before this unimaginable catastrophe, marine life had shown a dazzling splendour of forms quite unlike those of today but equally varied, forming richly complex food webs that had been evolving for over 100 million years. On land too, there were emerald forests bustling with insects and teeming with life. Some of this fecundity I have tried to capture in the preceding pages. Yet the end of this whole era of ancient life was near, these scenes of intriguing beauty and vitality from the Urals, Australia and El Capitan, and from many other places

FIG. 10 Photographic detail of marine trilobite fossils typical of the late Permian Tethys Ocean – note that one of the smaller trilobites has been fossilized still curled-up in a protective position, width of view 15 cm (Photo by Claire Ashford).

besides, simply faded into oblivion and were never to be reconstructed in exactly the same way again.

Solitary corals, sea lilies, graptolites and encrusting bryozoans mostly died out. Nearly all of the 160 known species of brachiopods perished. The last of the trilobites crawled over the seabed – a once invincible army of arthropods that had marched through the changing seas for 300 million years (Fig. 10). In the transparent upper waters, legions of plankton suffered irreparably, and these oceanic gardens of tiny foraminiferans and radiolarians that had sustained life at the base of the food chain were all but lost. On land, the great coniferous forests of Laurasia and Gondwana were wiped out, as was the distinctive *Glossopteris* flora, taking with them a host of insects and other small creatures. For the first and only time recorded in their long and varied fossil history, even the insects were knocked sideways – eight whole orders became extinct.

Equally difficult to explain for any extinction event are the groups that survive such catastrophes. In this case, the sharks and many other fish, as well as bottom-dwelling invertebrates (sea snails, bivalves and benthic foraminiferans), somehow sailed on seemingly unscathed. Some types of terrestrial vegetation made it through, and a very few of the larger mammal-like reptiles and amphibians managed to survive.

This great mass extinction event was quite unprecedented in scale and effect, and still its true duration and cause remain tantalizingly elusive. Part of the reason for this lies in the lack of good, continuous and fossil-bearing sequences across the Permian – Triassic boundary almost anywhere in the world. No rocks of this age can be drilled beneath the oceans, for the sea-floor everywhere is much younger. On land, there is either a major break in the rock record – a hiatus in time – at exactly this period, or there are probably continuous desert-lain deposits of red-coloured, cross-bedded sandstones that yield no fossils at all. Only in the Salt Mountains of Pakistan and, most recently, in Greenland has the event been found to have been recorded in its entirety. Partly because of the lack of data, it has generally been considered as a long, drawn-out affair, perhaps lasting for 8–10 million years. Certainly, a number of forms were in decline for a long period before the end – but not all. The latest studies from Greenland, however, suggest that there was also a geologically more rapid demise of many species over 80,000 years.

Two major environmental factors conspired with dramatic effect. The first was the slow but immensely influential fusing together of continents during the Permian period to form the supercontinent Pangaea. The consequences were manifold, although the process was very slow. Sea level fell to an all-time low, its lowest point in the whole 542 million years of the Phanerozoic Era. It was around 250 metres lower than at the beginning of the Permian Period and 300 metres below that to which it would rise again during the Great Flood at the acme of Tethyan dominance. The seas simply rolled back from the land until only 13% of the continental shelf areas were submerged, thereby greatly reducing habitat diversity and promoting ecological instability. Shallow shelf seas are regarded ecologically as the nurseries and hothouses of the ocean world – their loss must have had terrible consequences. I will discuss some of the reasons for sea-level change through time when considering the Great Flood in Chapter 6. At the same time, continental climates were extreme, severely stressing plant and animal life on land.

A further important result of lowered sea level was the exposure of many of the peat-rich deltas and swamplands that were a legacy of the great Carboniferous and Permian forests. The rapid oxidation of these richly carbonaceous deposits when exposed to the air released large quantities of carbon dioxide into the atmosphere and, by normal processes of air–sea exchange, transferred much of this into the ocean waters. Global warming and changes in ocean chemistry were an inevitable result.

The second and relatively more rapid environmental event was a very intense period of volcanic activity, for which we have good evidence within and to the east of the Ural Mountains.

Here the 'Siberian Traps' (step-like layers of successive lava flows) represent enormous outpourings of basalt coupled with highly explosive volcanism whose deposits now cover an area of 2–3 million square kilometres. In terms of the mantle plumes and hotspot volcanism implicated in the rupture and break-up of Pangaea (Chapter 2), this was a superplume event and the largest of its kind for which we have evidence in the rock record. It is not easy to date the beginning and end of this event; certainly it lasted for less than one million years and perhaps very much less. Immense clouds of volcanic ash may have temporarily blotted out sunlight and led to a 'nuclear winter'. But ultimately the result of vast emissions of carbon dioxide, sulphur dioxide and other greenhouse gases led to global warming, and probably also to poisoning of the atmosphere and oceans with fluorine, acid rain and trace metals.

In summary, I would argue that fundamental and dramatic changes in ocean chemistry, no doubt induced by the various environmental factors outlined above, were the ultimate cause of Permian extinction in the oceans. Such changes were catastrophic for life at the base of the oceanic food chain – and the rest, so to speak, is history.

We have probably taken the story as far as we can with present evidence and research. Pushed to their limits by such a combination of environmental factors, the natural decline of species increased many times over. The loss of certain plants on land, algae and other plankton at sea, affected a host of animals that fed on them and, in turn, the predators further up the food chain. Competition inevitably increased as favourable habitats diminished. Diseases may have spread more aggressively between animal groups as once isolated landmasses came together. In the final analysis, it was an unparalleled biological disaster from which the world took many, many millennia to recover. The broad and benign waters that characterized the early Tethys and harboured such a rich diversity of life were among the hardest hit. The early Triassic seas of the Tethys Ocean that followed, lapping at the dawn of a new era, were a shadow of their former grandeur, while the continents mostly remained hot and barren. It was to be almost 5 million years before plant life on the continents returned to anything like its erstwhile glory. It appears, however, that the warm inviting waters of the Tethys Ocean recovered a little more quickly.

PRESENT-DAY EXTINCTIONS

Before leaving the gloomy but intriguing topic of this chapter, let us briefly consider some parallels between the PT event and the world today. Indeed, if we begin to compare the present with past extinctions, then the result is quite frightening in its implications. Recent estimates for extinction rates during the 1990s fall between 1,000 and 10,000 species per year. This compares alarmingly with an average background rate of 1–10 species per year, and is believed to be close to the maximum rate witnessed during the great Permian extinction. Within the next three decades, that figure could have increased more than tenfold.

Most of this decline is as a direct or indirect result of human activity. Many larger animals have been hunted to extinction, while others die out as land is cleared for farming and development depriving them of their natural homes. The wholesale deforestation of temperate lands through the past millennium is now being followed by catastrophic destruction of tropical rainforests. The loss in biodiversity, much of which we have still not discovered, is incalculable. Human activity wreaks untold pollution across the globe, so that once-safe habitats on land and at sea are slowly being destroyed. The effects on global warming of burning fossil fuels are now well documented. Importantly in this regard, climate change has been one of the single biggest factors implicated in past mass extinctions. Many species today are already poised at dangerous levels for survival, just as they were in and surrounding the Tethys Ocean in the late Permian. Ocean acidification, as a direct result of increased levels of atmospheric carbon dioxide being transferred to the sea, is beginning to seriously and adversely affect the marine realm. Was it the same for Tethys at the end of the Permian?

There seems little doubt that we are currently living through a mass extinction event of our own making, which may yet prove even more severe than the PT event was for the Tethyan world. And yet the world today, on land and in the oceans, presents an unrivalled spectacle of rich diversity. This is an apparent conundrum. Perhaps resilient life is fast spreading back to colonize the fouled nest, and we will only gradually notice the change. Or, maybe, many of the species the world is losing are those we have not yet discovered, and the worst effects of our human excesses are yet to come.

4

Tethyan Fecundity in the Jurassic Seas

Near Lyme
beyond the shadowed sundial swing
lies Jurassic time

dormant fossils torn
by tides' claws
from secretive sea cliffs

Black Ven
like a smuggler
hides his ammonites

compressed life
beyond our knowing
curled in a coma of stone

From *Jurassic Time* by Peneli

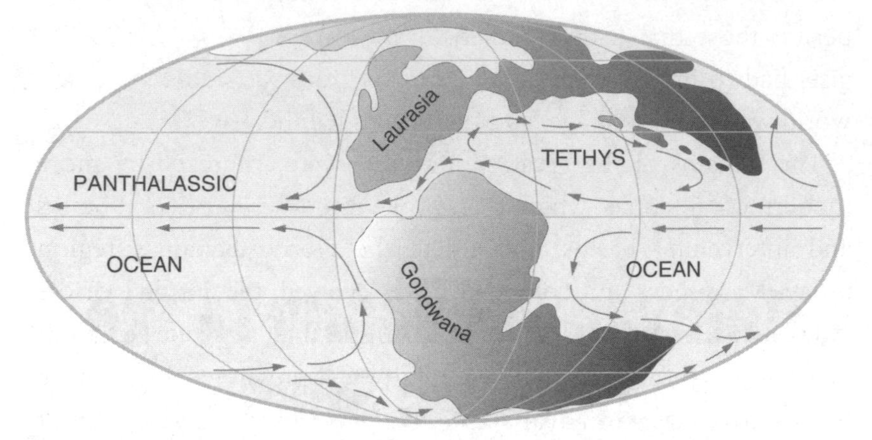

Mid Jurassic Tethys map (175 Mya). Global reconstruction with oceanic circulation, also showing an arm of the Tethys Ocean having broken westwards through Pangaea. The single supercontinent had divided into two: Laurasia to the north and Gondwana to the south.

After such a near catastrophe for life on Earth, there came a new beginning. The warm shallows that surrounded the Tethys Ocean, stripped bare of their former glory, became ideal nursery grounds for a completely new and varied flora and fauna that was set to evolve into some of the richest and most diverse ever. From primary producers at the base of the food chain to the top predators at its peak, new forms appeared and flourished in the Tethys Ocean. A deep-water sponge reef spread across the shelf seas of the northern Tethys, oasis to myriad colourful forms. And somewhere, probably in the swampy marshes at the ocean shoreline beyond these reefal lagoons, primitive crocodile-like *Thecodonts* evolved into the first true dinosaurs. For this was to become known as the Age of Dinosaurs (or the Age of Reptiles). Somewhere else, with less conspicuous fanfare, a small shrew-sized animal known as *Adelobasileus* became the first true mammal – but the Age of Mammals was still far from dawning. The plants and animals that grew and flourished were fundamentally different from those that came before, those that had characterized the Palaeozoic, so that geologists had to invent a new era. 'Mesozoic' means 'middle life' – the world was coming of age, and the oceans led the way.

The remains of the great sponge reef were colonized by more modern coral faunas, which then broke into a series of coral banks and intervening basins not unlike the Florida–Bahamas region today. Many new and colourful fishes evolved, playing live-or-die hide-and-seek with the sharks of old. It is in the fine white sediment of these coral lagoons that incredibly well-preserved fossils have been recovered from Bavaria and southern China, marking one of the most remarkable developments of the time – one that was to change the skies above land and sea forever – the evolution of flight. All this was yet to come.

VIRGIN OCEAN

It is hard to imagine the world, and the ocean world in particular, with such paucity of life as there was at the start of the Triassic, especially after the colourful variety of flora and fauna that we had come to expect in the latter stages of the Palaeozoic Era. The few opportunistic species that did survive not only came to flourish for a while in their own particular ecological niche, but also ventured out to colonize those left vacant after the PT extinction. Certain bottom-dwelling bivalves and sea snails multiplied and spread unchallenged across the sea-floor. The small and unprepossessing bivalve *Posidonia* with its rather flattened shells was one of those that came to dominate and, in places, its fossilized remains make up the bulk of the rock itself. In the waters above, the fish species that survived were also quick to flourish in numbers and to invade parts of the oceans they had hitherto avoided. Sharks did likewise. For a few millions of years in the earliest part of the Triassic, it seems that the numbers of single species far outweighed diversity. Of course this would not last. There was too much virgin territory to explore and new ecological niches to exploit; this was true for both the oceans and the land. It was only a matter of time before evolution had her way.

A combination of factors – suture of a supercontinent, unsurpassed violence of a superplume, and a chemical cocktail of change in the oceans – had conspired without showing the least sign of clemency and brought the world to its knees. It is as well to pause momentarily in this ultimate nadir, not least because current rates of extinction suggest close parallels between our own time and the Permian extinction, as outlined in the previous chapter, and we should not wish to visit that place again.

Perhaps there will always be a silver lining to each dark cloud that passes. Although Pangaea was still whole, at least for a short while, there were signs of rising sea levels and climatic amelioration. The Siberian superplume had spent its force and the atmosphere regained its composure. The ocean chemistry, though still with higher levels of carbon dioxide and less oxygen than before, at least remained relatively stable. The great Tethys Ocean was to become that silver lining with a new chemistry, new environmental niches and an abundance of ecospace vacated by the hapless passage of so many species at once. In short, it was warm, inviting and relatively uninhabited – an almost virgin ocean with precisely the ingredients for a radiation in life that would change the world forever.

In many respects, the next 40 million years of our story (the Triassic period) appears to be one of trial and error. Advances were followed by setbacks, radiation followed by extinction. But the principal templates for change were firmly established during this period. First to reassert itself was, of necessity, life at the base of the food chain, and indeed a whole new suite of phytoplankton and zooplankton appeared at the surface of those early Triassic Tethyan waters on which all else would come to depend. Sea level began to rise and marine waters spread across the formerly exposed continental shelves. Many new types of bivalve and sea snail evolved to take advantage of the near deserted sea-floor or to ease aside those monospecific opportunistic organisms that had so recently multiplied in numbers. Oysters appeared for the first time, together with a host of burrowing bivalves that lived in the sand and filtered small food particles from the seawater above. Limpets, periwinkles and top shells colonized rocky shorelines, crawling slowly across the surface of rocks and boulders to graze on a diet of algae and lichen. They were joined by two groups of animals that appeared for the first time in the Triassic and have since marched on through time, flourishing still in today's seas. These were the lobsters, shrimps and their close relatives, the *Decapoda* (having five pairs of legs on the thorax), and the scleractinian corals, which would soon become the supreme builders of framework reefs throughout the tropical world.

Swimming in the open waters overhead, it was the coiled shelled ammonites that first began to radiate significantly at this time. About a quarter of the multitude of known species of ammonite appeared in the Triassic seas. These were followed closely by a variety of marine reptiles. The southern shores of Tethys were patrolled by *Nothosaurs*, huge lizard-like reptiles with extended necks and long sharp teeth used for catching fish. *Placodonts* were particularly large ungainly reptiles with strong broad teeth for crushing tough-shelled oysters and limpets. They also hugged the coastline wading in and out of the water. *Ichthyosaurs* evolved their streamlined (dolphin-like) form very early in the Triassic Tethys and were soon to become one of the most cosmopolitan and well adapted of all.

But, for many of these marine organisms, either their initial evolutionary experiments came to nothing, or they were biding their time before they truly began to flourish in the Jurassic seas.

Other defining moments in evolution were being played out along the shores of the Tethys, in lowland swamps and along the course of rivers that swept an ever-increasing sediment load towards the ocean. The rivers were becoming muddier and floods more frequent because, very slowly, the climate across Pangaea was becoming wetter. Gradually, the extremes of desert aridity that I had seen in the reddened, wind-blasted sandstones of Morocco and beneath the North Sea were giving way to Triassic red beds of fluvial (riverine) origin. The mountains that had been pushed up as the supercontinent had sutured together continued to erode. The apron of alluvial fans, known as a *bajada* from the Spanish word, that flanked these mountain ranges, fed some of the largest rivers the world has known. Their sedimentary record, often following seamlessly in the rock record from the Permian desert sands, is all part of the New Red Sandstone.

I have walked across these distinctive red beds on almost every continent, following along the course of many an ancient river. From Brazil to Tanzania, from Arizona to Australia, from northern Europe to the Far East. Those of county Devon, my birthplace in southwest England, which cast imposing cliffs from Exmouth to Budleigh Salterton and form the westernmost part of England's World Heritage Coast, are some of the best preserved anywhere in the world. I am always captivated by their rich carmine colours and by the sweeping curves of lamination, the cross-bedding that records the power and direction of river flow from which they formed. Frequently the beds also contain fossil fragments of the plants that grew along the river banks and, ever since my experience in South Korea, I am on the lookout for something still more exciting, for footprints, for fossil bones, and perhaps more.

It was the summer of 2004 and I was working with my long-time colleague from my university days in Canada, Seung-Kwun Chough, who is now an eminent Professor of Sedimentology at Seoul National University in South Korea. In fact, we were carrying out some detailed work on a little-known sediment type that is deposited in front of high-flood rivers whose sediment-charged load forms a dense underflow down a steep submarine slope. They are called *hyperpycnites* and may indeed have formed in front of some of Pangaea's Triassic rivers although, to my knowledge, they have not yet been described from that era, and we were then looking at much younger deposits. It was hot and steamy work amid bamboo thickets and along the banks of insect-infested streams, so we had decided to take the day off and sample the culinary delights of Korean seafood on the coast west from Seoul. Now it often happens that a geologist's 'day off' turns into looking in a rather more leisurely way at rocks other than those being specifically being studied at the time.

The rocks we chose that afternoon were Triassic red beds. I could see immediately that we were dealing with small but powerful rivers, probably quite close to their mountain source and flowing south into the eastern reaches of Tethys. They may have been ephemeral in nature, with intermittent flash floods and longer periods of relative dryness. The conglomerates and sandstones contained numerous irregular-shaped pebbles of all different sizes, and there, suddenly, was something quite different – a few perfectly rounded 'pebbles' nestled together, some a bit cracked with the passage of time, and some with their outer layer partly peeled off. Close by we found another small group, and then another, and several other rounded 'pebbles' scattered individually. It turned out, after confirmation from paleontologists at Seoul National University, that we had discovered dinosaur eggs – a dinosaurs' nesting ground! I was too excited to notice that my camera was inadvertently set on video mode as I tried to photograph each of the nests, so to this day I have the whole event, including our delighted exclamations recorded live.

The most significant thing to note here is the evolution of the waterproof amniote egg in which the young were able to develop wholly within their own self-contained pond. The amniote egg contains a yolk that nourishes the developing embryo, as well as somewhere to store waste products, all while it is still protected within a hard or leathery outer shell. Just inside the shell is the amniotic membrane, which contains the vital fluids but enables gases to pass through. These important new attributes of egg structure and function mark the evolution of reptiles from amphibians and their ability, therefore, to live entirely on dry land without going through an aquatic larval stage. In fact, this evolutionary breakthrough had been achieved some 100 million years before, but most of the radiation from that time into primitive (anapsid) reptiles had been summarily cut short by the great Permian extinction. It was the diapsid reptiles that set about their amazing radiation through the Mesozoic Era whose eggs we found in that ancient stream bed in South Korea. Their descendants became the dinosaurs, swimming reptiles and flying reptiles of the Jurassic Period, as well as crocodiles, lizards and snakes that have survived to the present.

Before leaving the Pangaean continent and returning to the seas, I must make mention of a further evolutionary development in the plant world, which had, like almost everything else, been decimated at the end of the Permian. This was the development of gymnosperms (literally 'naked seeds'), where the seeds are exposed, often held in cones, and pollen is distributed on the wind. As with the amniote eggs of reptiles, their origin was before the Permian extinction event and was better suited to a relatively dry environment, but their true radiation came in the Mesozoic and, ultimately, paved the way for the evolution of flowering plants, or angiosperms. Initially it was the club mosses and seed ferns that started out on this brave new gymnosperm world, but they soon made way for forests of cycads, conifers and ginkgoes. Clearly, the adaptation was beneficial for all have closely living relatives today – some one hundred species of cycad in tropical forests, a single species of ginkgo, the maidenhair tree, and a great number of conifers, among which forms very similar to redwoods, Norfolk Island pines, cypresses and yews are all found as fossils in red-bed sediments of this age.

One of the most famous and quite spectacular occurrences of Triassic plants comes from the Petrified Forest in Arizona. Large fossilized trunks of monkey puzzle trees (*Araucaria*), scattered across the dry barren landscape, appear almost identical to modern types. The fossilization process with the very hard and durable mineral silica (quartz) has proceeded almost cell by cell, so that perfect detail of growth rings, external bark, knots, and even internal structures have been preserved. Not far away are found the first true dinosaur fossils dating from around 235 million years ago.

The evolution of dinosaurs and other reptiles was closely aligned with that of plants. The catastrophic demise of reptiles at the end-Permian event had hit the herbivorous reptiles hardest, so that only their meat-eating mammal-like cousins survived. When new floras appeared in the Triassic so a new group of herbivorous reptiles appeared and spread rapidly to all parts of the world. These were the rynchosaurs, a group of pig-like reptiles with broad jaws and stout tusks for digging up vegetation.

But the end of the Triassic was marked by another mass extinction event in which some 90% of land plant species died out, taking with them one-third of all land-based vertebrates, including the unfortunate rynchosaurs. The oceans also suffered a major setback as one-third of sea animals disappeared at this time. Many of the bottom-dwelling bivalves and gastropods, brave colonizers of the new Tethyan sea-floor, were wiped out, together with a host of new echinoid and crinoid species. The great majority of ammonoids disappeared almost as suddenly as they had come, with only very few genera surviving to ensure their even greater and renewed radiation at the beginning of the Jurassic Period that ensued. Marine reptiles also suffered greatly, including nothosaurs and placodonts, both of which had seemed so well suited to their new environment. Indeed, although it was almost as potent in its shake-up of life globally, as was the KT extinction event that finally wiped out the dinosaurs, very much less is known about what happened and why, at the end of the Triassic. There is little glamour, perhaps, in the death of almost all land plants and the animals that depended on them, in the demise of some but not all marine reptiles and in the near extinction of ammonites. Equally significant perhaps is the generally poorer preservation of fossils in the rock record in the latter part of the Triassic Period.

TETHYS STRIKES WEST

Pangaea was now beginning to break up and the Tethys Ocean to expand. Those 'short' 40 million years of Earth history during the Triassic were among the most enigmatic in the planet's long life. At the end of the Permian Period the whole of Pangaea found itself under great tensional stress so that deep fractures appeared quite dramatically and suddenly throughout the continent. As I mentioned in Chapter 2, this may have been related in some way to the enormous weight and large extent of the supercontinent causing overheating of the mantle below. This would have been especially true where it drifted over existing hotspots or even led to their development from hotspots to superplumes. But the truth is that we still do not have a good plate tectonic explanation for what was happening to Pangaea and to the Earth at this time. During the Triassic Period, these deep fractures cut clean through the continental crust and then typically developed into continental rifts. These were to have a profound effect on the shape of the world to come. They became the sites of thick sediment accumulation – of red beds and evaporites, giving way to marine deposits as the Tethys Ocean flooded in. At a much later time, it would be these sedimentary basins that became major oil provinces – the North Sea and Niger Delta are both founded on Triassic rift systems. They were also often the sites for outpourings of lava from underlying hotspots. But it is their role in the breaking of a supercontinent, the spread of the Tethys and, later still, in the opening up of the North and South Atlantic Oceans, that has been most significant of all.

I have travelled to many of these ancestral rift valleys around the world and, whether it is red beds, volcanic lavas or oil-rich sediments that I am examining, there are always the deeper, hidden questions – why is the rift here? And why did it start to form then? I will have to leave these more fundamental questions posed but not answered, and examine now some of the evidence of their nature and fill.

The near vertical cliffs of Palisades National park along the New Jersey banks of the Hudson River are made of dark grey basaltic rocks, turning rusty brown in places, that was intruded into the country rocks along one of these Triassic rifts. This type of intrusion is known as a 'sill' and is typically the feeding vent for lava flows nearby. The rocks were quite simply breathtaking in their tree-clad autumn colours when I last visited, but all the more so for the realization that I was walking along one of the first attempts at breaking up a supercontinent.

Palisades Sill is part of the extensive Newark Rift system, which stretches over 1000 kilometres from the Carolinas to Nova Scotia. Like so many of the other rifts from this time, it failed to develop further – fortunately, perhaps, as otherwise Manhattan Island and New York might have ended up next to Casablanca in Morocco! Interestingly, there is another failed rift on the opposite side of the

Atlantic – the Ziz Valley in Morocco. It has since been exhumed and raised to its present elevated position by uplift of the High Atlas Mountains, exposing the sediments formerly laid down on its floor. These include salt deposits and sun-baked mud flats, indicative of evaporation from a cut-off arm of the Tethys, followed by fully marine limestones with gigantic fossil ammonites, sponge mounds and algal reefs. Clearly the sea had flooded in and deepened still further to allow turbidite sediments to deposit, which are characteristic of deep-water environments. But the Ziz Valley too failed to become an ocean.

The rift that did finally open up sufficiently, such that sea-floor spreading began, ran through the Straits of Gibraltar and then cut southwest along the axis of what has since become the Central Atlantic Ocean. The Tethys Ocean flooded in from the east, past Morocco and Nova Scotia and reached as far south as Florida and the Bahamas. Ever so slowly the northern landmass, Laurasia, was inching apart from the southern continent of Gondwana.

Meanwhile, similarly complex rifting cut a gash between the Americas (or what were later to become North and South America) and the sea spilled in from Panthalassa in the west, more or less across the region that now runs from Yucatan along the Panamanian isthmus. Spreading began to form a proto-Gulf-of-Mexico but then stopped. These early seas dried up and left behind the telltale salt deposits that characterize so many early rifts. There must have been many episodes of flooding followed by evaporation, for the salt deposits in the Gulf of Mexico became very thick indeed. It is interesting to note here that much, much later, when these salt evaporites were buried under hundreds and thousands of metres of sediment eroded from the adjacent continent, they began to deform and flow under the great weight from above. Being of lower density and therefore buoyant, they pierced upwards through the overlying layers of sediment in finger-like protrusions known as salt domes. In some cases the salt forces its way upwards through several kilometres of overlying sediments and flows out over the sea floor. Indeed the present-day sea-floor of the northern Gulf is pockmarked with a maze of domes and hollows as a direct result of this bizarre behaviour of salt under pressure. It is a common feature of buried evaporites the world over, and readily visible on seismic profiles, both of which we shall return to in a later chapter.

Although spreading in the Gulf ceased, it was soon taken up to the south of the Yucatan Peninsula and a proto-Caribbean opened up. Finally, rifting and spreading managed to carve a passageway around the south of Florida and Cuba so that the waters pushing at either side of the divide could meet and the two great oceans of Panthalassa and Tethys were joined together. No longer was there an equatorial barrier to the free flow of water between oceans; no longer was there just one supercontinent. After all the rifting, tension and posturing, the dismemberment of Pangaea was a reality. Thus was opened the Tethys Seaway as a western extension of the Tethys Ocean – but a narrow seaway that was slowly becoming wider and wider.

It is salutary to pause here and let the enormity of what happened truly sink in. For just as the fusion of separate landmasses to form the Pangaean supercontinent is strongly implicated as one of the key causal factors in the greatest extinction event of all time (Chapter 3) so, I believe, its dismemberment must be held responsible for contributing to one of the principal episodes of evolutionary radiation that the world has ever seen. Taken together, the variety, profusion and fecundity of the Jurassic and Cretaceous seas is more than a match for that of any period of geological history, with the exception of the latter part of the Tertiary to Recent era – say the last 30 million years. Knowing that preservation of organic remains as fossils is a highly fragmentary and partial affair, it is clear that the geological record can yield only a sparse account of past life. Given this, it is quite possible that the Mesozoic Tethys at its heyday was more bountiful even than the richest oceans of today. The continents, with their luxuriant Jurassic forests and fearsome reign of the reptiles, might equally run our own rich tropics a close second.

Whatever the actual numbers of families and species that have existed at different times, which we will always have to estimate, and based on the best accountancy we have to date, the rate of radiation in the Mesozoic certainly equates with both that of the Cambrian explosion at the start of the Palaeozoic Era and that of the Tertiary explosion following from the death of the dinosaurs.

But why should the simple rifting and splintering apart of a continent be so instrumental in the radiation that was to follow? Continents have always been on the move about Earth's surface – what was so different about this event? These are very pertinent questions, but addressing them in full would cause me to stray too far from my central plot. Let it suffice to make a few brief comments here about the likely effects that the new shape of the Tethys Ocean would have as it came to wrap the planet in a girdle about its equatorial region.

First, and perhaps counter-intuitively, the unzipping of Pangaea coupled with new areas of active sea-floor spreading led to a global rise in sea level and the further spreading of seas across once exposed areas of continental shelf. This is because when new oceanic ridges rise up from the sea-floor they displace water already in the ocean thereby pushing sea level higher and drowning what was once low-lying land. Second, the penetration of an ocean, albeit a narrow seaway in the first instance, across the heart of what had been the most arid and formidable continental interior ever seen, brought about an immediate change in climate. There was more moisture available to the continents and their coastal regions, and an amelioration of conditions worldwide.

Third, and very significantly, this meagre oceanic connection single-handedly changed the pattern of oceanic circulation. For the first time in perhaps 150 million years, there was no barrier to flow from east to west around the equatorial world and, as the ocean deepened further, bottom currents as well as surface currents could utilize this oceanic gateway. The preceding configuration of Tethys forced circulation into two great gyres, thereby drawing warmer water away from the equator and returning with cooler water from higher latitudes. In this way the climatic extremes due to latitude are ameliorated. But with the supercontinent breached, the warm waters could simply flow around and around the equatorial ocean. Higher latitudes became partially isolated – not completely at this stage – and hence assumed a cooler climate.

Fourth, the creation of two continents rather than one (to be followed rapidly by still further divisions) greatly increased the length of coastline and, aided by higher levels of rainfall, continental erosion increased the contribution of sediment and minerals to the oceans. With a longer coastline, flooded shelf areas (from rising sea level) and more climatic variability, the availability and variability of ecospace increased. There were simply more areas of different ambient conditions for life to exploit. These factors also increased the number and distribution of regions of oceanic upwelling and nutrient-rich surface waters, while greater rates of erosion fed more mineral nutrients (especially iron) to the sea. Iron, in particular, is an essential ingredient for enhanced primary productivity in the oceans.

Fifth, continental separation immediately led to the geographical isolation of species and their progressive evolutionary divergence on the separated landmasses. Equally, there was the freedom to diversify, at least initially, without torrid competition between all animals together on a single continent. Exactly the same is true for the many shallow marine species for which a deep and wide ocean is a prohibitive barrier to migration. It is from this time that the northern and southern margins of the Tethys begin to assume distinctly different faunal assemblages.

Collectively, this combination of factors provided greater ecospace for adaptive radiation to proceed, an abundance of new ecological niches to allow experimentation and diversity, and improved levels of primary productivity to supply the base of the food chain both at sea and on land. It is little wonder then that the Jurassic seas were fecund almost to the point of promiscuity! For the rest of this chapter, I will turn my attention to charting the remarkable growth and vibrancy of the ocean world while dinosaurs ruled the land. This is a complementary story to the much better known, and more often told, tales of the dinosaurs.

THE JURASSIC SEA WORLD

It is testament to the profusion and significance of fossils from this period that the Jurassic system was the first stratigraphic unit to be formally mapped and classified, even before the science of stratigraphy had been invented. William Smith, the English engineer charged with building a system of canals across southern and central England in the late 18th century, made two crucial observations during this work. He noticed that the different sedimentary layers being excavated contained different sets of fossils. Furthermore, these distinctive fossil assemblages were repeated wherever the same particular layer was encountered. He introduced what is now known as the 'law of faunal succession', a concept later used by Darwin in support of his theory of evolution, and that has remained as one of the cornerstones of stratigraphy ever since. He further compiled the first geological maps, representing the outcrops of different ages (based on their fossil assemblages) with different colours on the map. All this work was carried out in Jurassic strata.

At about the same time, the Swiss geologist Alexander von Humboldt was working on rocks of the same age and general style in the Jura Mountains of northern Switzerland. He referred to them as the 'Jura Kalkstein' (Jura Limestone) in his publication of 1795, from which the formal name of the Jurassic system has since been derived. Others in Europe encountered very similar sedimentary rocks with the same fossil assemblages – in France, Spain, Italy, Germany, Austria, Poland and more. Certainly, the Tethyan seas of the Jurassic spread far and wide across this part of Europe, and so the rocks they deposited, with the fossils they preserve, have yielded many of the early and important clues for reconstructing the sea world at that time.

Let me start at the base of the food chain. There can be no mistake: then as now, phytoplankton – the photosynthesizing component of plankton – was the single most important support for the whole of ocean life. This largely invisible microscopic world was made up of countless billions of single-celled organisms, floaters and drifters of the surface waters, which basked in energizing sunlight and were bathed with the inorganic nutrients present in all seawater. They had all learnt to perform a relatively simple chemical reaction – the process of photosynthesis whereby carbon dioxide and water combine to give a sugar called glucose. Glucose itself is a useful food and source of energy, but it is also essential in the construction of more complex carbohydrates. Oxygen is a by-product of the reaction, expelled into the hydrosphere and atmosphere thereby allowing the majority of other organisms to breathe. A special green-coloured pigment known as chlorophyll absorbs energy from sunlight in order to fuel this all-important chemical reaction.

Collectively, such organisms are known as autotrophs because they manufacture their own food. They are the primary producers that form the solid base of nearly all food chains in the ocean, and are therefore many times more abundant than the organisms that feed on them – known as heterotrophs. All primary producers of food are rapidly grazed by an ever-burgeoning army of hungry herbivores. In their turn the herbivores are eaten by carnivores. These are, respectively, the primary and secondary consumers within the ocean ecosystem.

In those early Jurassic seas, new types of microscopic primary producers took hold of the phytoplankton world. Coccoliths and, later to appear in the fossil record, diatoms bloomed in countless abundance in the clear surface waters, building their tiny lime and silica skeletons in a beautiful and varied array of intricate shapes. Dinoflagellates, the principal primary producers in warm waters today, also became common at this time. There was, of course, a parallel explosive radiation of zooplankton (the animal component of plankton) to feed upon such bounty – lime-secreting foraminifers at the forefront, siliceous radiolarians not far behind. All these we will meet again in Chapter 6, when they reached their absolute acme. But just what creatures higher up the food chain took advantage of such a rich brew?

For the amateur fossil collector, sea-floor sediments typically abound with the remains of organisms that scavenged on this renewed plenty (Fig. 11). Marine invertebrates capitalized on the ocean rejuvenation that was now under way with an exuberant radiation in form. Molluscs proliferated – from the oysters and clams that hugged, burrowed or attached themselves to the substrate, and great reef-like mounds constructed almost entirely by a new type of thick-shelled bivalve (known as a rudist), to the highly effective free-swimming shellfish (cephalopods). New types of corals and bryozoans appeared in warm shallow seas. And everywhere the sea-floor was terrorized by voracious predators – crabs and lobsters able to crack open even

FIG. 11 Jurassic marine diorama based on Tethyan fossils: 1 Archaeopteryx – the first known bird, 2 Medusoid jellyfish and surface plankton, 3 Saccocoma – a floating crinoid, 4 and 5 Early Teleost (modern) fishes, 6 Ammonite, 7 Belemnite, 8 Horseshoe crab, 9 Bryozoans, 10 and 11 Hexacorals (Scleractinian) building up a 'modern' coral reef on an algal-encrusted sponge reef (12) substrate.

the hardest of shells, and meat-eating snails. Burrowing became a common avoidance technique.

Most coveted and also most abundant as fossils of this time are a magnificent variety of coiled ammonites, some growing to as much as a metre in diameter (Fig. 12a), and bullet-shaped belemnites, related to modern squid. Ammonites take their name from their resemblance to the horns of the Egyptian god Ammon, actually the Greek name for an oracle god whose sanctuary was situated at Siwa oasis in the heart of the Libyan desert. Libyan desert tribes first worshipped Amun, a god in the shape of a ram, but following its adoption and passage through several ancient civilizations, the name and meaning changed slightly. The Greek word 'ammos' actually means sand.

Mythology aside, the ammonites were a highly successful group of molluscs, known as cephalopods, all of which are now extinct apart from the squid, octopus and nautiloids. These modern relatives all propel themselves along with jets of water, and have well-developed eyes, brains and nervous systems. The nautiloids have a strong shell as defensive armour against shell-crushing predators, and chambers filled with variable proportions of liquid and gas for buoyancy. We assume that ammonites possessed similar attributes, were generally good swimmers (at least those with more streamlined shells), and occupied the same niches as fish today. Literally thousands of different species have been described, indicative of their rapid evolution over a short time span and their divergence to fill many different ecological niches (as do fish today). The smallest were no larger than a euro coin (or United States quarter); the largest about the size of a wheel from a forty-ton truck! Together with the belemnites, these formed a colourful, cosmopolitan population of predators and scavengers across the bountiful seas.

THE WORLD HERITAGE JURASSIC COAST

It is not hard to know just where to travel to experience at first hand the wonder of this ancient seaworld. The Jura Mountains are spectacular indeed, parts of the Tethys sea-floor that were caught up much later in that final phase of mountain building, which brought the ocean to such a dramatic end as our story closes (Chapter 10). But they are more for the mountaineer or serious walker than for most of us. The central Apennines in the charming Italian provinces of Umbria and Marches are far more accessible, and there has been some excellent work by Italian paleontologists and sedimentologists in unravelling the history of this region. Indeed, among them is one of my former PhD students, Dr Manuela Marconi, who painstakingly demonstrated from the sediment record the existence of a series of banks and basins along this northern portion of the Tethys. This kind of sea-floor topography was ideal for inducing local upwelling of cooler nutrient-rich waters from below and thereby stimulating a biological orgy in the surface waters. But in this part of the world the rock outcrops are too often covered by magnificent fields of sunflowers or vineyards.

North Africa and the Middle East also boast some of the best Jurassic outcrops anywhere, which formerly lay along the southern margin of the Tethys. Many of the ammonites sold to University Geology Departments for their teaching collection or found in fossil shops and museums, come from Morocco. In my travels to the Atlas mountains I have tried on many occasions to elicit the whereabouts of some of these rich fossil localities from the Berber roadside salesmen but, not surprisingly, they have never yet revealed their secrets.

But for the finest example of Jurassic rocks, where else should one go, of course, other than the World Heritage Jurassic Coast of Dorset and East Devon in southern England? It is the first region of natural landscape representing a geological system to be awarded World Heritage Site status – an honour received in December 2001. William Arkell, who authored *Jurassic Geology of the World* in 1956, had earlier written of the Dorset coastline: 'Few tracts of equal size could raise so many claims, scientific, aesthetic and literary, for preservation . . . ' He followed in the footsteps of many eminent geologists of the 19th and early 20th centuries who visited or worked along this hallowed coastline, and for whom it became known as a crucible of knowledge in this most fashionable of scientific pursuits. To this day it remains an outdoor classroom and training ground for tens of thousands of students every year, as well as a delight for millions of other visitors. I was first introduced to it myself by two colleagues, both formerly of Southampton University: Professor Michael House, whose detailed research on the rhythmic limestone cycles at Lyme Regis linked their origin to astronomically induced cyclic changes in Earth's climate; and Dr Ian West, whose breadth of knowledge of the region taught me much, both scientific and historical, and is elegantly captured and illustrated in his personal website.

Since then I have led numerous geological excursions along these shores myself, for students and colleagues from such far flung places as Japan, South Korea, Brazil, Nigeria and North America. The intrigue and delight of new discoveries and new perspectives remain as fresh as ever, just as the tranquil scenery and coastal footpaths never cease to charm. The images wrought from the rocks of that Jurassic past could fill a book and more (as they have from the pen and lens of many before me), so it is not easy to select the best nuggets for my story now. Let me mention just a few.

Early Jurassic life is best represented in the rocks near Lyme Regis and Charmouth, where it was first dramatically exposed by an amateur fossil collector from Dorset, Mary Anning, in the first half

FIG. 12 Photographic detail of marine fossils typical of the Jurassic Tethys Ocean: (A) Ammonites – width of view 15 cm (Photo by Claire Ashford); (B) Echinoids (sea urchins), coral, and bivlaves – width of view 25 cm (Photo by Claire Ashford); (C) Part of Ichtyosaur jaw bones and teeth – width of view 20cm (Photo by Claire Ashford).

FIG. 12 (Continued).

of the 19th century. Her collection and documentation of marine reptiles was second to none, with many 'first-finds' and exceptionally well-preserved, near-complete skeletons. The picture that emerged was one from the very top of the food chain – saltwater crocodiles, big-headed short-necked pliosaurs, and paddle-finned long-necked plesiosaurs. All had strong jaws with numerous sharp teeth, and some grew to as much as 12 metres in length. Perhaps most highly evolved were the streamlined, dolphin-like ichthyosaurs, reptiles that first returned to the sea in the Triassic period and, by this stage, had even abandoned the amniote egg in favour of giving birth to live young (as recorded by some remarkable fossil finds). Mary Anning uncovered a host of ammonites and other marine fossils (Fig. 12), and even examples of flying reptiles, insects and dinosaur bones that had been caught up in the lime muds of those shallow coastal seas and lagoons. The rocks at Lyme are all from what has since been formally adopted as the *Liassic* stage of the Jurassic system, named from the Dorset quarryman's term 'lias' meaning 'layers of rock'.

Because of her pioneering work, Mary Anning – a poor Dorset woman with no formal education – became guide and companion to many of the most eminent geologists of her time. Mary was only once enticed out of her home county to present her work in London in 1829, by Sir Roderick Murchison, then President of the Geological and Royal Geographical Societies. But she did become good friends with Sir Henry de la Beche, founder of the British Geological Survey, whose inspired diorama *Duria Antiquior*, painted in 1930 and based on Mary's collection, was the first published scientific reconstruction of a past world (Fig. 13). This portrayal of the Jurassic seaworld is every bit as full of intrigue, excitement and danger as that of the dinosaur domination on land.

It is an unstable coastline and with each new rock fall more of that hidden past is revealed. Almost 200 years later, another amateur fossil collector, Kevan Sheehan, has pieced together the largest

FIG. 13 The first published scientific reconstruction of a past world – Duria Antiquior (a more ancient Dorsetshire) – drawn by Sir Henry de la Beche in 1830, inspired by the Tethys fossil finds of Mary Anning. (© Oxford University Museum of Natural History/The Bridgeman Art Library)

complete skull of a pliosaur yet discovered. Although the rest of the creature is almost certainly still buried beneath the crumbling cliff, we can estimate its length at around 17 metres – about twice as long as a double-decker bus. Its powerful jaws and gigantic teeth could have bitten a small car clean in two, or taken a *Tyrannosaurus rex* in one gulp, had one ever strayed into those shallow seas for a cooling bath. Undoubtedly, these Tethyan marine reptiles were the true masters of Jurassic time.

The *Middle Jurassic* can be examined along the coast at Osmington Mills, just down from the Smuggler's Inn, a secluded cliff-top pub dating from the 17th century that hides an intriguing history of shipwrecks and looting, as well as fine pie and ale after the day's exploration is done. The section captures an interesting environmental change from river-fed sands and muds of a storm-pounded shelf and shoreface (the signs of which are clearly visible in the fine sedimentary structures preserved), giving way to sediment-free seawater from which pristine white calcium carbonate has precipitated out as tiny spherical sand grains in warm agitated lagoons. These distinctive sediments, known as *oolites*, are a well-known feature of Jurassic limestones in other parts of Europe, as well as of other ages and places where similar conditions had prevailed.

Besides the numerous fossils of marine invertebrates, this stretch of coastline is one of the best I know for examining the tracks and trails, burrows, resting places and homes of a plethora of animals that once lived on and below the sea-floor. Collectively, they are known as trace fossils (or ichnofossils) and are studied by paleontological specialists known as ichnologists. Tantalizingly, the animals that made the tracks are rarely preserved as fossils themselves, although I have been lucky enough, while hunting with students along the Osmington foreshore, to find several small echinoid fossils (sea urchins) that appear to have become trapped at the end of their burrows. Mostly, however, we can learn much about how and where the animals of such traces must have lived from careful comparison with similar sea-floor features today, whose authors we can identify.

The *Upper Jurassic* is represented by some classic localities, also designated as the *type sections* internationally. Most important are the dark-grey coloured laminated shales with flattened layers of bivalve and ammonite fossils found at Kimmeridge Bay, and after which the *Kimmeridgian Stage* has been named. These shales (well-compacted mudstones) contain large quantities of organic matter, which is 'cooked' when deeply buried to form oil and gas. The same Kimmeridge shales lie deep beneath the North Sea where they have given rise to such enormous volumes of oil that a whole industry has been built up around its exploitation. What is important to note here is that, towards the end of the Jurassic period, a subtle change of environmental conditions led to widespread stagnation of parts of the sea-floor (over many parts of the Tethys world) coupled with enormous over-productivity of the plankton in the surface waters. This led to the preservation of organic matter in the sediment.

These unusual conditions did not last for long – just long enough to store away a billion dollar treasure chest deep within the rocks – before the world returned to normality once more. The next layers to be laid down in this region were lime-rich and shell-packed sediment that became cemented on burial to form a magnificent, sparkling grey-white limestone known as Portland Stone. This has been quarried since Roman times for its durable qualities and attractive appearance, and it has since become one of the most famous building stones in the world. It was chosen by the great architect and engineer, Sir Christopher Wren, to rebuild many fine buildings across the Capital after the Great Fire of London in 1666 – including St Paul's Cathedral. The first quarries and still the best today are found on the Isle of Portland just south of Weymouth, from which inestimable tons have since been removed. Portland stone is also known for some of the largest ammonites ever found, such as *Titanites*, whose coiled shell grew up to 60 centimetres in diameter, with lethal squid-like tentacles protruding an unknown length behind the last dwelling chamber. These magnificent ammonite fossils are often used locally as decorative additions to stone walls.

My final snapshot from the Jurassic Coast comes later still in time and is best seen near the outermost part of Lulworth Cove. Earth movements related to a later phase of Tethys closure have pushed these Jurassic rocks up above land and further tilted them almost vertically. The seaward line of cliffs are made of very hard Portland Stone, together with the overlying, almost equally durable, Purbeck limestone. Continued battering from the waves over interminable ages eventually forced a narrow breach, probably along a lineament in the rocks weakened by fractures related to their uplift and then cannibalized by a small stream draining to the coast. Once the sea gained access, it took much less time to carve out an almost perfectly circular bay in the softer (and younger, Cretaceous-age) rocks behind. This is Lulworth Cove.

During Purbeck time, the sea level dropped locally, although worldwide it was on the rise, and a series of islands emerged in the Dorset area, surrounded by saline lagoons on the Tethyan rim. Preserved in the rocks are fossil soils and the remains of a once luxuriant tropical forest of giant cypress, monkey puzzle trees, cycads and ferns. Stretching from Mupe Bay near Lulworth to Portland and Weymouth is the most complete fossil record of a Jurassic forest in the world. On the outer ledges at Lulworth, there is evidence of the sea returning and giant tree stumps being flooded under a shallow saline lagoon, such that they became covered with a mat of lime-secreting algae, which built up in layers over time to form large doughnut-shaped mounds known as stromatolites.

JURASSIC REEFS

No one who has ever snorkelled or dived on a coral reef in the warm tropical shallows of a Pacific or Indian Ocean atoll can fail to be mesmerized by the sheer abundance and audacity of the life they harbour. Thousands of unique and colourful, harmless or menacing species coexist in one of the ocean's most flamboyant spectacles. And so it has been for over 500 million years since the first reef-like mounds were constructed by lime-secreting algae and bacteria on the Cambrian shelves and seamounts. Reef communities developed and proliferated – they offered food in abundance, a veritable labyrinth of hollows and chambers for security or surprise, and a well-honed dating service for finding a mate. The Silurian, Devonian and Carboniferous reefs were each in their own way magnificent and unique constructions – their fragmented remains now form some of the world's most arresting limestone scenery, while polished slabs of fossil packed stone are widely used throughout the construction industry. The Permian-age reef of El Capitan in Texas (Chapter 3) revealed something of the spectacle that life had become just before catastrophic extinction descended as a dark shadow over this Palaeozoic world.

Changing reef environments through time have doubtless spawned a plethora of new species. Although we can never trace a single mutation to its actual birthplace, it is very probable that fledgling fish first tried out their new fins and jaws in that colourful underworld of the Silurian seas some 425 million years ago. They would have found ample protection from larger predators within the shelter of sea lilies, corals and stony stromatolite hideaways. Once the true fish had mastered the art of swimming, these gymnasts of the submarine world were set to conquer the seas. This they achieved with great purpose and alacrity swimming, literally, to the far corners of the Earth, rapidly displacing their well-armoured predecessors – the jawless fishes. They continued to adapt and evolve through the Mesozoic seas of the Tethys, at times doubtless rivalling the colourful pageant they present today.

Primitive reef-like structures of lime-secreting algae and hardy bivalves must have been among the early invaders at the Tethys shoreline, beginning a brave new world after widespread nemesis had wiped clean the slate. Clear warm lagoons formed along the edge of the sea in the protected environment behind such reefs and shell banks. A rejuvenated flora in swamplands and nascent forests spilled from the land. Safety is always of paramount importance to all creatures that live on or near the sea-floor. Perhaps it was in this setting that ammonites first evolved and grew to rival even the longer-lived fishes in their mastery at sea. Surely too, it was here that some reptiles slipped quietly back into the water and so began their ascent into fearful predators at the top of the food chain, as so elegantly unravelled by Mary Anning's work and evocatively portrayed in De La Beche's diorama.

Already then with an incredible diversity of intriguing new life forms, the Tethys world was very much alive and evolving. A deep-water sponge reef spread across the northern shelf seas, the remains of which can be found scattered in rocks that stretch from what is now southern Spain all the way through France, Switzerland, Germany and Poland to the Romanian Black Sea coast. This distance is estimated at nearly 3000 kilometres, one-and-a-half times as long as the Great Barrier Reef off northeast Australia today. Giant clams and bizarre thick-walled rudists helped bind the reef framework, magnificent coiled ammonites and squid-like belemnites explored its flanks together with myriad fish and giant sea turtles (newcomers to Mesozoic seas), while voracious crabs, lobsters and meat-eating snails prowled the sea-floor. But few parts were safe from the attention of fast-swimming ichthyosaurs, powerful pliosaurs and the ever-present threat from flying reptiles (pterodactyls) in the skies above.

This impressive sponge reef, almost certainly the largest such feature yet to decorate Tethys waters, was completely hidden from the surface, anchored on the gentle slope of the outer continental shelf at around 150 metres water depth. A somewhat similar and very extensive deep-water reef system has been recently discovered offshore western Ireland and Scotland today, although this lies deeper still on the continental slope. Tectonic earth movements along the north Tethys margin at this time led to progressive shallowing of the seas, such that the great sponge reef came too close to the surface for comfort. As large tracts died off, the dead skeletal remains were colonized by newly evolved species of hexacoral – similar to the typical scleractinian corals that dominate today.

So efficient it seems were these new builders of framework reefs, or perhaps the conditions were just right for them to exploit, that they managed to completely cut off a very large lagoonal area, over 2000 square kilometres in size. Very fine limy debris periodically washed over from the broken reef crest during storms, but otherwise the waters of the lagoon lay quiet and undisturbed. High rates of evaporation led to extreme salinities and a low oxygen content, the lagoon became a toxic bath for any who dared venture near, and the fine sediments that accumulated layer by layer over the next half million years embalmed and preserved in fantastic detail any organisms that fell to the floor of the lagoon. These extraordinary conditions are recorded in the limestone rocks today that outcrop in a region of Bavaria in southern Germany around the small town of

Solnhofen. Occurrences such as the Solnhofen limestone, with such exceptional preservation of fossils, are very rare through the enormity of geological time but, where they do occur, they provide a unique window on past life. It was the serendipitous discovery in these rocks, while quarrying for building stone in 1861, of the first fossil bird *Archaeopteryx*, that provided a key milestone in our understanding of the evolution of birds from their reptilian ancestors.

ESCAPE TO THE AIR

In order to fly, an animal must resemble a small aircraft. It must be light in weight, possess strong, covered wings, sport a fully aerodynamic shape and be powered by an adequate fuel source. Birds have all these characteristics – hollow bones, feathered wings, streamlined shape, powerful muscles and heart, and a high metabolic rate to supply fuel for flight. They are both warm-blooded and endothermic (i.e. producing their own heat). The other major animal group to have taken wing are the insects. Because of their extremely light weight and smaller size, they do not require a warm-blooded metabolism for fuel maintenance. But what about attempts at flight in the fossil record? How did evolution mastermind an escape to the air?

As far as we know the first to achieve this were the insects near the start of the Carboniferous period some 350 million years ago, a development that probably went hand in hand with the evolution of terrestrial plants. Flight clearly brought many advantages to insects – escape from predators, locating new sources of food, and finding safe places to mate and lay eggs – and the great new forests that developed at the end of the Palaeozoic Era would have been ideal for such new experimentation and adaptation. Four main theories have been advanced for the evolution of wings in insects: allowing for safer drifting in strong gusts of wind; better using the wind for sailing over the surface of water; increasing the length of jump as a means of locomotion and escape; and gliding efficiently from branch to branch and stem to stem. It is quite possible that more than one of these routes led to successful flight.

The second group of animals to take to the air were the reptiles, and these evolved in two distinct lineages – the pterosaurs and the thero-pod dinosaurs. Both lineages diverged from archosaur reptiles in the Triassic. The same sorts of underlying advantages would have driven vertebrates to achieve flight, as with the much smaller invertebrate insects. The principal evolutionary drivers that led the way are believed to have been: an ability to glide down from trees, rocks or cliffs and swoop on unsuspecting prey below; a better ability to leap between branches on trees; or take off from land after running fast, either as a means of escape or for chase. The evolution of both lines may have run in parallel for a while but, ultimately, only one was to prevail.

Pterosaurs quickly became diverse and prolific, ranging far and wide. Some were as small as sparrows, others as large as hang-gliders with unbelievably broad and elegant wingspans up to 15 metres across. While the smaller pterosaurs were doubtless wing flappers, the larger ones were giant gliders, soaring on thermal updrafts much as vultures and eagles do today. Somewhere in between in size were those more akin to seabirds of today, acrobatic fliers able to dive, snatch and grab for fish or other delicacies in the shallow seas. Others lived further inland and may have fed mainly on a diet of insects. In all cases, their bones were hollow, very light and fragile as in birds, but their wings were stretched with a thin, tough, mem-branous material, rather than feathers, that linked together front and hind legs. This characteristic, in particular, suggests that they were not an evolutionary line for birds.

The first true birds are represented by *Archaeopteryx* (meaning 'ancient wing') from the Solnhofen lagoon, with superb preservation revealing the detailed structure of feathered wings and a single tail feather. These creatures date from about 150 million years ago – the late Jurassic Period. Interestingly, evidence now suggests that they evolved from small, actively running, bipedal, theropod dinosaurs, and not from the pterosaurs that had already taken to the air. They were closely related to the *Velociraptors,* or 'raptors' of *Jurassic Park* fame. Although only six specimens of *Archaeopteryx* have yet been found, other feathered theropods and primitive birds, both from this period and from later in the Cretaceous, are beginning to appear in other areas surrounding the Tethys Ocean, most notably the aptly named *Confuciusornis* from north-east China.

Much speculation continues as to just why and where the evolutionary advantages that led to birds really took hold. But I have included discussion here because, for me, there are many good reasons for suggesting that the changing Tethys Ocean at that time was strongly instrumental in this momentous advance. Certainly, the fossils found to date have all been from lagoonal sediments of Tethys affinity. More significantly there was a new and rich supply of food to be had from among the fish at sea, the shellfish along the shoreline, and the many burrowers of annelid or other affinity on intertidal flats. The extended coastline and rising sea level gave rise to a plethora of estuaries, inlets, bays and lagoons – the nursery grounds for marine life and relatively safe places for nesting. Reptiles of all kinds were no strangers to the Tethys shore for the same sorts of reasons that I have listed above. Some were waders of that in-between world, some became wholly aquatic, and others evolved into flying reptiles – the pterosaurs. This indeed would have been the obvious place with the key drivers for birds to evolve.

5

Black Death to Black Gold

A lingering day was enveloped by water,
by fire, by smoke, by silence, by gold,
by silver, by ashes, by passing and there
it lay scattered, the longest of days:
the tree tumbled whole and calcified,
one century then another hid it away
until a broad slab of stone forever
replaced the rustling of its leaves...

From *Stones of the Sky* by Pablo Neruda
(translated by James Nolan)

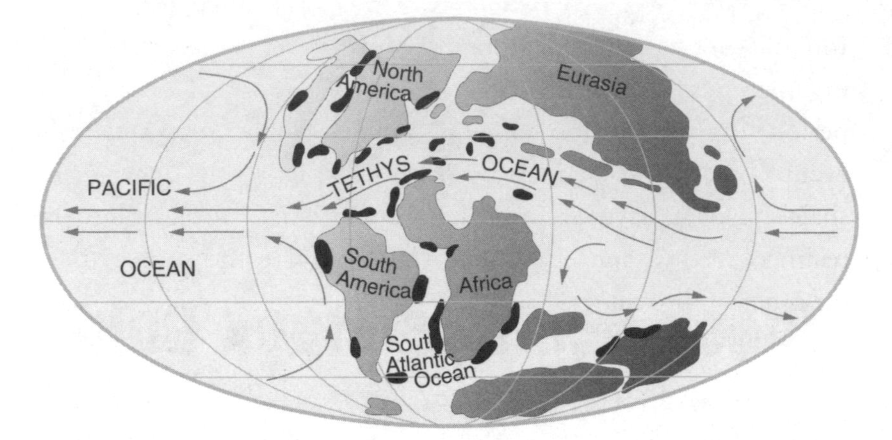

Mid Cretaceous Tethys map (95 Mya). Global reconstruction with oceanic circulation, showing the Tethys as a broad ocean separating the dispersed remnants of Laurasia and Gondwana. The solid black shows areas of black shale accumulation between 120 and 90 Mya.

Life in this vibrant ocean was neither always nor everywhere benign. The erstwhile blue and well aerated seas could temporarily turn into a dark nightmare of deep stagnant waters from which there was no escape. As the planktonic world continued to flourish in the warm surface waters, at times with an almost manic over-productivity, a crisis in oxygen levels far below awaited the constant rain of billions upon billions of microscopic organisms after their brief lives were over. The deep putrid waters and seafloor muds preserved even the soft organic tissues of these tiny life forms from further decay in astounding abundance. Their remains coloured the sediment black.

Remarkably, it is the oceanic remains of the Mesozoic Era – preserved in sediments known as black shales, deeply buried and cooked by the Earth's internal heat engine – that are the source of between 60–70% of the world's oil. This is true for the whole Middle East region, with its bountiful fields of black gold, which lay on the southern Tethys shores at the time of the Black Death of

100 million years ago. All around the former Tethys there are black organic rocks of exactly this age, testament to an ocean-wide period of severe oxygen depletion – what geologists term an anoxic event – which was perhaps even global in its reach. Other black shale episodes have occurred before and since, both within the realm of Tethys and elsewhere. These provide the source rocks for other oil provinces.

CHARTING THE BLACK DEATH AT SEA

I came across the Black Death very early in my career as a geologist while I was still working for Britoil. Somewhat unexpectedly, and probably as a result of papers I had recently published on deep-water sedimentation, I was offered the opportunity of sailing on an international scientific expedition to the South Atlantic Ocean with the Deep Sea Drilling Program (DSDP) in 1980. Not hesitating to accept, I thought little more about the science involved, apart from some preplanning of potential post-cruise research on samples we were to collect from deep beneath the sea-floor. We met our colleagues and shipmates-to-be in New York, a diverse group of scientists who had converged from the UK, France, Germany, Austria, Japan and the USA – mostly experts in our different specialized fields. On the extremely long charter plane trek to join the ship at Walvis Bay in Namibia, via Rio de Janeiro and Windhoek, we were able to explore our individual interests in rather more depth. I soon learnt that several colleagues – organic geochemists – were there solely for the black shales they expected to encounter when we drilled into sediments of Middle Cretaceous age. My curiosity and expectation were duly aroused. What were these sediments and why so significant?

The *Glomar Challenger* was a giant of a ship compared with those I had sailed on before, much more like the North Sea oil rigs that were currently a part of my job. The drilling derrick towered 50 metres above the drill floor and straddled the moon pool, a rectangular hole in the midship through which drilling took place. There were six tiers of scientific laboratories, a library, meeting rooms, and even computer suites, which was highly advanced for those early days of the revolution in information technology. We steamed for five days out into the South Atlantic, somewhere north of the Tropic of Capricorn in what is called the Angola Basin, reaching into the grey-blue depths of that mysterious and remote loneliness that I have only ever experienced in true open ocean. There we began to drill.

The drilling process itself is a combination of brute strength and technological wizardry. In a repetitive and physically demanding process of great precision, ten metre lengths of heavy steel drill pipe are hoisted vertically inside the derrick, mechanically screwed onto the top of the drill pipes that have already been assembled and hang below ship, and then the whole drill string is lowered a further ten metres into the yawning abyss... and so the process continues. Twenty-four hours a day, seven days a week, to the accompaniment of loud music and much liquid intake by the drill crew. The tropical sun was scorching, but the sea breeze provided a welcome respite. Drilling in 5560 metres water depth meant that 556 sections of drill pipe had to be assembled before we had even plumbed the void and reached sea-floor. A hugely heavy, flexible, spaghetti-like, steel strand was suspended below us, buffeted by currents, nuzzled by inquisitive fishes, and poised to core into the completely unknown, untouched and unseen landscape of the deep.

The sense of scientific adventure and exploration was palpable as the first core came on deck. And thereafter too, for each new section drilled took us still further back into deep time – approximately ten metres per one million years of history at the drill site we had chosen (DSDP Site 530). Every ten metres had its own story to tell. We saw the start of the Benguela cold current and nutrient-rich upwelling system at five million years ago, which has made this offshore region so rich for fisheries today. There were gigantic submarine landslides and debris flows that had swept down slope ten million years ago, probably triggering a monstrous tsunami in the ocean waters above. Deeper still, we drilled clean across the KT boundary (the dinosaur extinction event), where we witnessed little more than a gentle transition of species – more of which in Chapter 7. Still further back in time there had been an extended period of submarine volcanism, some of the volcanic seamounts growing above sea level on the Walvis Ridge, where they became fringed with coral reefs. And then . . . and then there was the Black Death.

At around 850 metres below the sea-floor we drilled into our first black, organic-rich sediments; layer upon layer, some thick some thin, intercalated with lighter-coloured greyish or greenish sediment and extending over some 200 metres of section. This was a geochemist's delight and a most interesting story for us all. The shipboard micropalaeontologists quickly dated this interval to the mid Cretaceous, between 100 and 85 million years ago, on the basis of the species of microfossils that they separated out from the sediment. Indeed, one of the most rewarding aspects of such international expeditions, apart from the gems of discovery that drilling into virgin sea-floor inevitably yields, is to have a team of some 25 scientists each bringing their own experience and specialism to bear on the same scientific challenges. In this case, having dated what I refer to, scientifically, as a 'black shale episode', the challenge was to unravel some of the mystery behind this most widespread and notorious of all such episodes that have occurred either before or since. Certainly there was much we could do aboard ship, and we each took samples of the precious shales to analyse in our separate laboratories on *Glomar Challenger*, but the Black Death had a much more global reach and significance that required still more intense post-cruise scrutiny.

TETHYAN REACH

That was Leg Number 75 of the Deep Sea Drilling Project, part of an inspired and extended scientific enquiry into the ocean history of planet Earth, a programme which has been running more or less continuously since 1968. The DSDP started almost a century after the world's first major oceanographic expedition, which covered some 70,000 nautical miles between 1872 and 1876, aboard *HMS Challenger*. Sadly, neither the original ship nor her namesake *Glomar Challenger* still ply the waves, although I was invited to sail on her final leg in 1983 on which we drilled in the deep Gulf of Mexico towards the western extremity of our former Tethys Ocean. She was thereafter replaced by *Joides Resolution* for what became the Ocean Drilling Program (ODP), and has most recently been joined by an even grander and more capable drillship *Chikyu*, launching a new phase known as the Integrated Ocean Drilling Program (IODP).

I have intentionally digressed briefly into this historical perspective for two reasons. First, it is from this point of (geological) time forwards that I can begin to use the collected results from many hundreds of holes drilled into the ocean floor since the scientific drilling programme began, in order to more accurately chart something of Tethyan history. Ocean floor older than the Cretaceous has very rarely been drilled; the oldest recorded is of mid-Jurassic age from the far western Pacific, which was then part of Panthalassa. This is because, as it grows older, ocean crust is continually being subducted back into the mantle through the submarine trenches that occur at convergent plate margins. As new ocean crust moves ever further away from the Mid-Ocean Ridge where it was formed it becomes progressively cooler and denser and ultimately more prone to subduction. Second, I will start by looking at some of these other sites for further evidence of the mid-Cretaceous black-shale episode.

My interrogation began at sea, as I buried myself amid weighty tomes of past DSDP results in the *Glomar Challenger* library, in between my 12-hour shifts in the sediment lab and relaxing spells of table tennis in the gym or more serious running round the helipad (33 circuits to the kilometre, if I remember rightly). And, of course, it has continued since as more and more deep ocean sites have been drilled. What quickly became apparent is that there were two main black-shale episodes in the mid-Cretaceous, which both occur almost everywhere within reach of Tethys at that time. The events are clearly recorded in deep holes beneath the Mediterranean and Black Seas, the Central Atlantic and the Caribbean, but are also present beneath much of the present-day North Atlantic and South Atlantic, where we drilled Site 530 for example. Fingers from the Tethys extended along these north–south rifts, nascent oceans that were just beginning to open up.

We drilled two further sites during the leg, the last of these perched on the Walvis Ridge at the outer edge of the present day Benguela upwelling system, for which we had already seen evidence at Site 530. It gave me more food for thought, in terms of just how and where organic matter can accumulate in the sediments of today's oceans, during the long 'homeward' voyage clear across the South Atlantic to Recife on the north-east coast of Brazil.

Not long afterwards I found myself based in central Italy at what is surely one of the most attractive medieval university towns anywhere. At the heart of the Marche region, Urbino is a small walled city perched on a steep-sided hill – a lively and picturesque UNESCO World Heritage Site, noted for its fine historical legacy of Renaissance art and culture. The university was founded in 1506, but has recently been completely redesigned by the architect Giancarlo de Carlo, so that modern offices and lecture theatres are hidden within an unchanged outer facade of the medieval city, with its tortuous streets and chaotic piazzas. His aim was to create an urban tissue where students, staff and citizens would be fully integrated. Whether or not these grandiose ambitions have been realized in Urbino, it was certainly a fine place to contemplate the black shale mystery and to stroll with colleagues at dusk (that most refined of Italian customs) in the Piazza del Duca Federico, after a hard day in the field.

The Marche and Umbrian regions of Italy provide some of the best localities in Europe to examine the mid-Cretaceous black-shale episodes on land. Here, however, the episode is recorded by black organic-rich chert beds. Chert is a hard quartz-rich rock – the same material as the flints in chalk cliffs or the pebbles on many beaches. The black bands of organic-rich chert are intercalated with white limestone – spanning exactly the same time periods as found in the DSDP and ODP boreholes beneath the ocean floor. This cyclicity in the rocks is yet another challenge for explanation. I have now seen the same black-shale episodes and their cycles of organic-rich/organic-poor beds wherever the rising Tethys of the time spilled over the continents – from the Black Sea coast in Ukraine, through Italy, Sicily, France, Spain, south-west United States and Mexico. Along the southern margin of the Tethys, the distribution of black shales is still more extensive, from north-west Australia through the Middle East, Morocco, Algeria and Tunisia, all the way to Venezuela in the far west.

In order to explain the widespread burial of organic carbon in marine sediments at this time (or, indeed, at any other time) we must consider two fundamental controls on the supply of large amounts of organic matter to the sea-floor and then on its preservation within the sediment. The first is primary biological productivity and recycling, and the second is how the oceans are stirred. To understand these controls and their effects is particularly important because neither excessive supply nor easy preservation of organic matter is the norm in the oceans today. I will first turn to what we know of modern oceans.

PRODUCTIVITY AND RECYCLING

As we saw in Chapter 4, the ocean surface is truly the garden of the world – a garden of sumptuous fertility, a plethora of creativity and perfectionism that at times seems almost a competition of ocean art. Each year over 6000 billion tons of phytoplankton grow wherever light penetrates the waters. These are all single-celled protists with tiny specks of chlorophyll, which harness solar energy to synthesize their tiny organic bodies from nutrient chemicals in seawater. It is a simple, efficient and extremely old process that was certainly as conspicuous in the Tethys Ocean as it is today. But primary productivity in the marine plankton is far from constant, either in time or space. When conditions are optimum for their rapid growth, then they reproduce in such phenomenal abundance that huge areas of sea are literally swarming with life – this is commonly referred to as a spring bloom.

Delving a little more deeply into this thriving garden, one of the most abundant groups are the *diatoms,* miniature organisms that make their homes in beautiful silica glass boxes of all shapes and designs. They live life in the fast lane, reproducing asexually every 12–24 hours until the resulting cells eventually become too small for viable subdivision. At this point they abandon old ways, cast aside their glass cases, and reproduce sexually, once more growing to their former size before secreting a new home. Most diatoms store food reserves as little droplets of oils and fatty acids, which also help maintain buoyancy and hence a prime position as they jostle for the best place to sunbathe.

Two other groups that especially love tropical seas and the open ocean are the *coccoliths* and *dinoflagellates*. Coccoliths build their complex spherical homes of tiny patterned discs carefully constructed from calcium carbonate (lime). At certain seasons, they can multiply at a phenomenal rate until a milky turquoise spring bloom incorporating a staggering 25 billion billion individuals can be observed on satellite photographs spreading across 250,000 square kilometres of ocean – an area the size of Great Britain. Dinoflagellates, by contrast, are commonly seen eerily glowing with a ghostly green-blue bioluminescence at night. They have a tough cellulose coating and are active swimmers among the plankton, aided by their two long whip-like flagella. But some species have abandoned cell walls for a comfortable symbiotic life within the bodies of certain jellyfish, corals and molluscs. They supply the food, while their host provides protection and nutrients. As far as we know, this lifestyle was equally in evidence in the Tethys.

Until recently, the *cyanobacteria* went unnoticed and unsampled in the phytoplankton, because of their ultra-microscopic size. More than half a million of their individual cells would fit onto a single pinhead, and yet they can account for a staggering 80% of primary production at the ocean surface. In shallow waters, where sunlight easily penetrates, they are also found carpeting the sea-floor as rather slippery, dense green mats. Those forms that secrete calcium carbonate are the remarkable architects of the weird and wonderful cabbage-like stromatolites – infamous long-term residents of Shark Bay in Western Australia. In this case, small is undoubtedly both beautiful and highly successful, for it is this same group of photosynthesizing bacteria that laid down the original template for life on Earth 3.5 billion years ago.

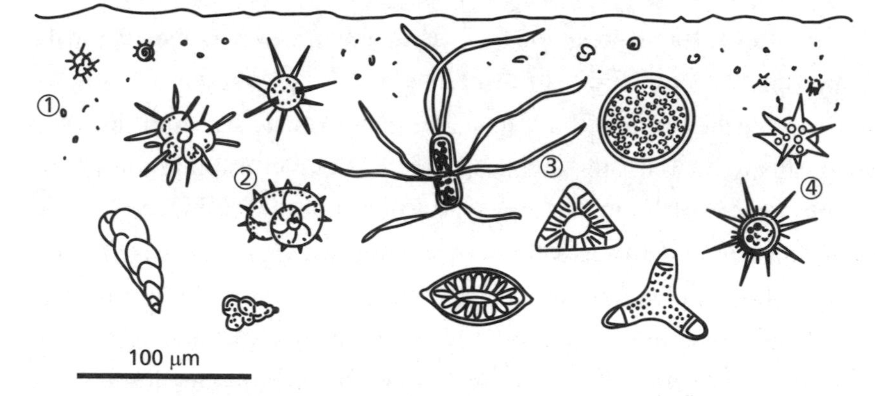

FIG. 14 Cretaceous fossil plankton diorama based on Tethyan microfossils: 1, Nannoplankton (coccolithophores) and juvenile foraminifers; 2, Foraminifers (five large, two very small – all have calcareous shells); 3, Diatoms (five large, one shown with protruding filaments – all have silica glass shells); 4, Radiolarians (two large with spicules – all have silica glass shells). Scale bar is 100 microns (0.1 mm) long.

All these organisms are members of the phytoplankton community – the primary producers par excellence (Fig. 14). They are accompanied in close pursuit by a voracious army of peaceful grazers and tiny predators – the zooplankton (Fig. 14). These are also single-celled organisms that can build their tiny sculptured shells, in this case about the size of sand grains, at extraordinary rates in order to take advantage of spring blooms in phytoplankton. There are elaborate multichambered shells of *foraminiferans,* made from calcium carbonate, and spiky silica glass spheres of *radiolarians.* Both types of organism extend fine radiating strands of their own cell material (filopedia – or 'feet of thread') through tiny pores in their shells, easily ensnaring other minute organisms, drawing the living material into the parent cell where it is chemically broken down by powerful enzymes. Even organisms of comparable size can be caught and broken up externally, and their dismembered body parts added to the feast.

In short, this Garden of Eden is a world of open warfare, with nowhere to hide and no prisoners taken live. Together the phytoplankton and zooplankton feed the rest of the ocean world, passing up through short but often complex food chains and webs, to the top marine predators of the day. In fact, life in the microscopic world is brief whether you are eaten or die naturally – from a few hours to a few weeks at most. If you die, you sink.

Everywhere in the oceans there is a constant, unremitting, downward drift of material from the surface towards the sea-floor. Dead and decaying organic matter, large and small faeces, furiously active bacteria, mineral skeletons cast by planktonic species in the vain hope of protection through life, and inorganic detritus windblown or washed in from the continents. Many of the individual particles are tiny and would take an interminable age to sink. But together they build larger aggregates of material – sticky or fragile, porous or dense, elaborate or simple – all falling as 'marine snow'. In particular parts of the ocean and at certain times of the year when surface productivity is high, the falls are thick and fast, reaching to even the deepest parts in a matter of weeks, and blanketing the sea-floor with a soft grey carpet of organic-rich mud.

At the first signs of death, a host of tiny microbes move in and take over. They are the undertakers and sewage workers, the waste collectors and recycling merchants of the ocean world. They live among the thriving plankton at the surface waiting for death or excrement; they hover in the twilight zone of the middle ocean to hitch a ride on falling debris; and they even lurk on the deepest sea-floor, that ever-sinister world without sunlight. Their role in death is vital for life – decomposing the remains of dead organisms and animal waste products, thereby releasing a wealth of essential nutrients back into the environment. Principal among the nutrients for life are carbon, nitrogen, phosphorus, and silicon. Bacteria are skilled specialists, each with a different role in the breakdown of complex organic molecules, working in close consort to effect almost complete recycling.

For the greatest part, therefore, today's oceans are well aerated and the waste products derived from such fantastic productivity, including carbon, are efficiently recycled. In certain regions, however, where productivity is excessively high, as in areas of upwelling offshore Peru or Namibia, or in stagnant waters greatly enriched by decaying organisms and their waste, such as the Black Sea or the Gulf of California, conditions are locally reversed. The organic matter supply exceeds the recycling capability of the microbial armies and the supply of free oxygen is used up. This is because the first wave of microbes use oxygen for their metabolism – they are aerobic. When they can't cope, their reinforcements use up still more oxygen. Once all the free oxygen has gone, other, non-oxygen-using or anaerobic bacteria take over. But these microbes decompose organic matter much more slowly, and part of the carbon is left behind in the sediment.

Such were the conditions of productivity and recycling in the Tethys Ocean during the mid-Cretaceous.

STIRRING THE OCEANS

The currents that stir the oceans are over 2000 times more powerful than any rivers on land (Fig. 15). Their flow is continuous, transporting a phenomenal quantity of water around the face of the planet in a gigantic circulatory system. It is the constant turmoil of the atmosphere, and the winds that are powered by a never-ending supply of solar energy, that keeps the ocean waters in perpetual motion. And the distribution of continental landmasses, intersecting the principal flows, determines their overall pattern. The Gulf Stream in the western North Atlantic, for example, transports over 55 million cubic metres of water every second – which is around 1000 times the total discharge rate of

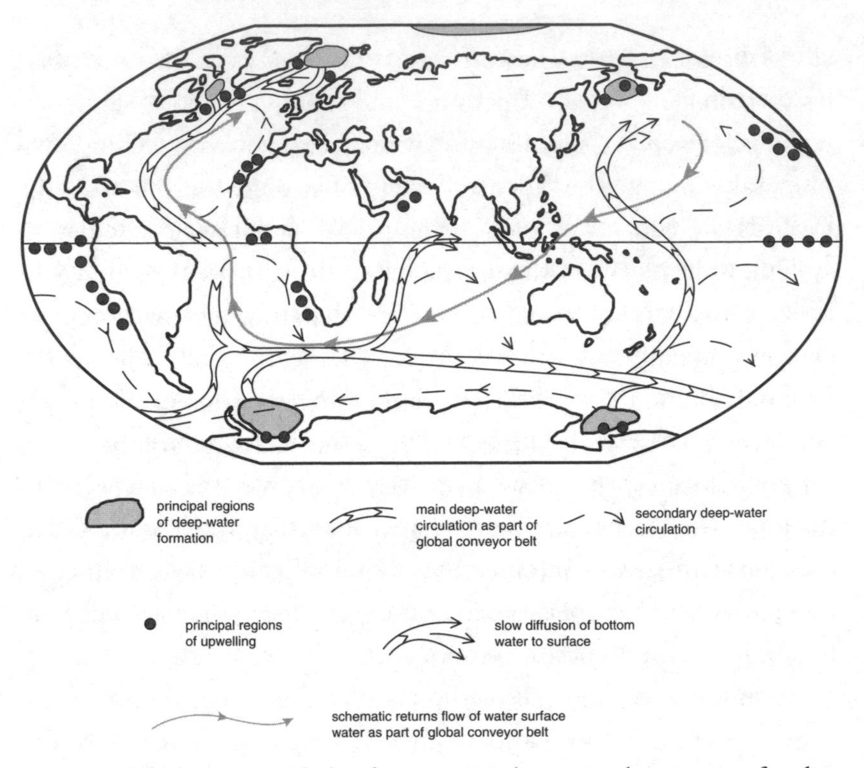

FIG. 15 Global conveyor belt of ocean circulation in the oceans of today, also showing the principal regions of deepwater formation and main areas of upwelling.

the world's top 20 rivers added together. The southward directed Canary Current is a broader, slower return flow along the eastern margin of the basin. Similar gyres circling north and south of the equator operate in each of the main oceans, as a result of the north – south orientation of continents and of the oceans in between. Only in the Southern Ocean that encircles Antarctica are the currents free to flow uninterrupted around the globe. This circulation in turn provides a very effective insulation of the polar region from equatorial waters.

Linked with the great surface gyres, but entirely hidden from view, are powerful, omnipresent, very slow-moving currents that flow along the ocean floor. They are driven by density differences caused by variations in water temperature and salinity. Ice cold, salt-laden water is formed at the poles as seawater freezes, leaving behind 70% of its salt content. These denser waters sink to the bottom and flow towards the equator as part of the global '*thermohaline*' circulation system. Collectively, both surface and bottom currents form a vast oceanic network of circulating water that transfers heat energy, carbon dioxide and other nutrient elements, and sedimentary material around the world. This is elegantly referred to as the 'global conveyer belt' by Wally Broecker, an eminent oceanographer from the United States. The principal driver for the global conveyer belt is the high temperature differential between the equator and the poles, coupled with seasonal sea-ice development. It has been estimated that one complete cycle, for any given water molecule, will take between 1000 and 1500 years.

We now know that the oceans play a crucial part in the global climate system. The immense amounts of heat and moisture stored within them serve both to moderate change and to prolong it once it commences. They also serve as a gigantic sponge for critical climate gases, holding 50 times more carbon dioxide than the atmosphere. Colder waters at high latitudes, and throughout the deep sea, hold more dissolved oxygen than low-latitude warmer waters. The ocean surface acts as a two-way control valve for gas exchange, opening and closing in response to two main factors: gas concentration and ocean stirring.

Given the present distribution of continents and the resultant ocean circulation, we are currently living in what is known as an *icehouse world*, albeit during a temporary warm respite from ice-age conditions. Average sea-surface temperatures range from 27°C at the equator to −1.5°C near Antarctica. Below 2000 metres depth, the deep oceans are bathed in uniformly cold waters of 1–4°C. Thick permanent ice caps cover polar region landmasses, while extensive sea-ice occurs seasonally. Global trends towards climate warming are beginning to temper these extremes, but on a scale that does not begin to approach the *greenhouse world* of Cretaceous time.

One further aspect of ocean circulation that is important to examine here is that of oceanic upwelling and the stirring of nutrients that are essential for primary productivity. The principal nutrient elements include carbon, nitrogen, phosphorus and silicon, together with certain trace metals, such as iron for chlorophyll and other pigments. They are supplied in abundance from rivers that empty billions of tonnes of chemicals and organic detritus annually into the oceans. They are also held in the living tissues of organisms and released back into seawater from the bacterial recycling of dead and decaying organisms and their faeces. But these organic materials have a tendency to sink rapidly to the sea-floor and take vital nutrients with them. Their return to the surface occurs most effectively in regions of upwelling, where cooler waters at several hundred metres depth are forced upwards, as well as by thorough mixing of the upper ocean layer during major winter storms.

Upwelling takes place where different water masses meet in the oceans and circulation allows deeper water to reach the surface. In today's oceans this occurs in one prominent equatorial and two high latitude belts. It is even more pronounced in certain coastal areas where prevailing surface currents are directed away from land so that deeper waters rise to fill the void. Particularly strong coastal upwelling of this sort occurs offshore West Africa, Chile – Peru and California (Fig. 15).

TOWARDS AN EXPLANATION

Lest you had forgotten, and my digression into present day oceans gives you every reason to have done so, I am building towards an explanation of the black-shale episodes of mid-Cretaceous – the Black

Death that was so pervasive across the breadth of Tethys and even beyond.

That the Cretaceous was indeed a greenhouse world is now abundantly clear. Tethys had opened as a broad low-latitude ocean providing an equatorial gateway for ocean circulation. Bathed in continuous warmth, the ocean currents that flowed through Tethys and circumnavigated the world were heated and continued to heat up. Gyres that peeled off to north and south spilled these warm waters towards the poles. Our plate tectonic reconstructions of the time show no polar landmasses, and high latitude marine sediments of that age have shown no evidence of floating ice, as in the Arctic Ocean today. We also find fossil forests in rocks from the Antarctic and from Canada that were then located around latitude 85° north and south (i.e. equivalent to the central Arctic Ocean today), coals that formed beyond 60° (i.e. north of Oslo in Norway) and coral reefs far beyond their subtropical distribution of today.

A further intriguing source of independent evidence for this greenhouse scenario comes from a very useful technique developed over the past few decades, pioneered by Sir Nicholas Shackleton of Cambridge University and now utilized by many scientists world-wide. This is based on the existence of two different isotopes of oxygen, oxygen-18 and oxygen-16, and the discovery that these occur in very slightly different proportions in the water molecules of colder versus warmer seas. Colder seawater is preferentially enriched in the heavy oxygen-18 isotope, and this same very subtle signal of enrichment is transferred to the chemical make-up of fossil shell material secreted by marine organisms. From the tiny skeletons of microscopic plankton to the thick-walled shells of oysters or giant clams on the sea-floor, all organisms that have used oxygen from seawater to build their protective hard parts of calcium carbonate, hold a record of the isotopic proportions of the oceans in which they lived. The data is incontrovertible: during the mid-Cretaceous, equatorial sea surface temperatures were 25–30°C, decreasing to a balmy 10–15°C at the poles; while 15°C has been recorded from fossil shells that lived at 2000 metres water depth.

Eric Barron, now the director of the National Centre for Atmospheric Sciences in Boulder, Colorado, was one of my colleagues aboard *Glomar Challenger* in the South Atlantic. He had first discovered these remarkable temperatures from fossils found in the Cretaceous sediments we cored in 1980. The crucial point about the temperature of the oceans is that warm water holds less oxygen. It escapes in gaseous form to the atmosphere or is used up by planktonic organisms in constructing their calcareous or siliceous skeletons.

Not only were the Cretaceous seas far warmer than those of today, but circulation would also have been far more sluggish without the key drivers of high temperature differential and seasonal sea-ice. Sea-level was considerably higher than today and still rising, so that broad areas of continents along the Tethys shoreline became flooded, and marginal seas and extensive lagoonal areas developed, all with a potential for restricted circulation and slow stagnation. The same almost certainly applied to fingers of Tethys that splayed into the still narrow South Atlantic, which was where my investigation of these black shales first began. Their warm waters held less oxygen. The sluggish circulation, partially restricted shelf areas and semi-enclosed marginal seas, would all have contributed to further oxygen reduction as the armies of microbes set about recycling dead organic matter, absorbing available oxygen in the process. Areas of oxygen starvation and complete stagnation were the inevitable result.

There is little doubt that primary productivity was high in the surface waters of the Tethys. Ample evidence is preserved in the rock record as the microfossil remains of countless billions of planktonic organisms that lived and died in the seas above. Perhaps the broad shelf areas would, in places, have made especially fertile gardens or factories for biogenic proliferation, and some of this organic largesse would have been transported into the deep ocean by major currents coursing down the continental slopes. This is exactly the situation we observed when drilling in the South Atlantic – organic debris from the Walvis Ridge being swept into the deep Angola basin along with fine clays and silts. When organic matter is buried rapidly, most bacteria and other scavengers can't easily reach it to decompose it. Furthermore, if it is deeply buried in fine clay-rich sediment then seawater carrying oxygen from above finds it almost impossible to percolate down. Good preservation is the result.

It is this combination of factors, culminating in low oceanic oxygen coupled with high organic matter supply, which led to such widespread and prolonged episodes of black shale accumulation in the Tethys Ocean between about 125 and 85 Mya. This is what my colleague from Oxford University, Professor Hugh Jenkyns, first referred to as an ocean anoxic event, and what I have called the Black Death at the start of the chapter.

THE STORY OF OIL

It is both a strange quirk of fate, as well as a hugely important scientific consequence, that the conditions leading to Black Death in the Tethys Ocean are exactly those required for the generation of large quantities of oil and gas, on which the very essence of 21st century life depends. At this point, therefore, I must delve briefly into the whole story of oil – how it forms and where it is trapped – and then consider the part played by Tethyan black shales in shaping our lives today.

Oil and gas, together with coal and peat, are fossil fuels based on the principal chemical elements common to all living matter – carbon and hydrogen. They are formed from the organic remains of dead plant material, through a process of decay and change over millions of years. Only a tiny fraction of the original living bulk is eventually preserved as a fuel resource. In the case of hydrocarbons (oil and gas), the story begins in sunlit surface waters of former seas and lakes, where billions upon billions of microscopic organisms make up the rich planktonic broth. Given specific environmental conditions, including high primary productivity, oxygen starvation and rapid burial in fine sediment, the resulting deposit is a black, organic-rich mud. This is exactly as I have just described above for the Cretaceous Black Death.

Within these black muds is the template for generation of the particular hydrocarbon molecules that make up oil or gas. Land plants, preserved in coal swamps or washed out to sea from deltas, tend to produce gas. The marine plankton, which contributes most to black shale, yields mainly oil, followed by gas at higher temperatures. The process by which organic matter matures, however, is immeasurably slow – like so much else that takes place in the geological realm. Slowly the muds are buried as more sediment is deposited above, slowly they become compressed and harden into black shales, and even more slowly the organic matter heats up.

There is a particular window of temperature, between about 50° and 100°C, through which the very complex organic molecules of life are broken down into long-chain hydrocarbons found in oil. At higher temperatures still (150–250°C), these are further cracked to yield the short-chain hydrocarbons, mainly methane, of natural gas. The typical thermal gradient, or rate of increase in temperature, is around 20–30°C per kilometre burial depth, beneath most parts of the ocean floor. This means that oil generation commonly occurs between 2 and 6 kilometres below the sea-floor, and gas at greater depths. It may take 10–20 million years of sediment accumulation before burial to these depths is achieved.

Widely dispersed oil or gas in fine-grained, compressed black shale is very difficult to extract, and almost impossible when it is buried deep below the sea-floor. However, the very great pressures that result from such deep burial force the mobile hydrocarbons, together with water still trapped in the sediment from the time of deposition, to migrate outwards and upwards. It is estimated that as much as 90% of the oil and gas originally generated leaks out at the surface in natural seeps and vents, part of the grand recycling of nutrient elements.

The remaining 10% is held back in the tiny pore spaces of thick accumulations of sandstone, and other sediments, or in minute cavities in some limestones, where shell debris and coral fragments have been partially dissolved. These are the reservoir rocks that oil explorationists seek out and can then exploit. They fill up with oil and gas rather like a bath sponge with water. They can be deposited in many different ways, in oceanic or continental environments, but need to be of sufficient size that they hold economically significant reserves. The reservoir geometry or shape deep within the subsurface must form a trap with a covering seal, into which hydrocarbons can migrate but from which there is no escape. Only then will the oil and gas continue to collect in these giant sponge pools below the surface (Fig. 16).

BLACK GOLD

The mid-Cretaceous black shales of the Tethys are well recognized as the most prolific source rock for oil in the world. What now comprise the dozen or more nation states of the Middle East, seat of OPEC – that select group of the world's principal Oil Producing

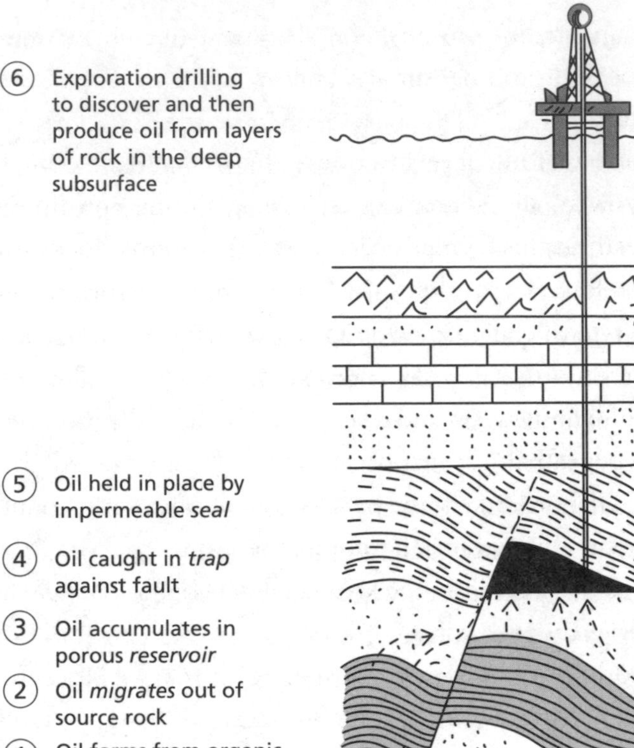

FIG. 16 Factors required for origin and accumulation of oil.

and Exporting Countries – are there but by an accident of geology, which placed them on the margins of the Tethys Ocean for much of their early history. They are part of a geological region known as the golden crescent, which now produces some two-thirds of the world's oil and over a quarter of its gas. Many of the giant and supergiant oilfields, as well as a host of smaller ones, have been fed directly by this Cretaceous oil. The same is true for many other prolific oil provinces, from the Sirte Basin in Libya to the

Venezuelan giants, from offshore Northwest Australia to those that simply ooze from the surface of the Earth in Kazakhstan and Uzbekistan.

All these regions have in common a Tethyan descent. In fact, not only were the source rocks formed during the time of the Black Death in this former ocean, but the reservoir rocks too were deposited somewhere along the shoreline or continental slope of Tethys – typically at a time somewhat later in its history. The close juxtaposition of source rock and reservoir is a significant bonus for the ease of oil migration and reservoir fill. One of the most recent oil booms is beneath the deeper ocean waters on either side of the South Atlantic, offshore Brazil and several of the opposing countries in West Africa. Once again, the dominant source rocks are the mid-Cretaceous black shales – the same ones that so caught my imagination in the Angola Basin.

The world has seen other black-shale episodes, both before and since, and several of these are important source rocks for other oil fields and provinces. Some I have already mentioned as having formed at different times, but still within Tethys waters. The Jurassic black shales that formed when Tethys waters first spread northwards across Europe, and that are now so well exposed along parts of the World Heritage Coast of southern England, are the principal source rocks for the combined North Sea and northwest European oil provinces. Others, such as the Palaeozoic source rocks across much of North Africa, were deposited long before the Tethys was born. Younger Cenozoic source rocks, such as those which feed the offshore oilfields in the Gulf of Mexico, are more properly assigned to oceans that have appeared since. In all cases, the story of oil generation from black shales is exactly the same. The coincidence of many different factors and their almost serendipitous alignment through geological time is what has created the fuel resource we most crave (Fig. 16). It is worth remarking here that humankind can burn in a matter of seconds what nature has taken countless millennia to produce – a reliable estimate is that, globally, we consume annually an amount of fossil fuel that it took one million years to form.

6

The Greatest Flood of All Time: Rise and Fall of the Seas

Lonely, majestic . . .
that imperial sweep of chalk cliffs
where the strident calls of gulls
fall as pale echoes
and broken flint husks lie abandoned
at the tide's frayed edge.
Only time's eternal silence
masks a microscopic cornucopia
of bygone seas.

From *Chalk Seas* by Dorrik Stow

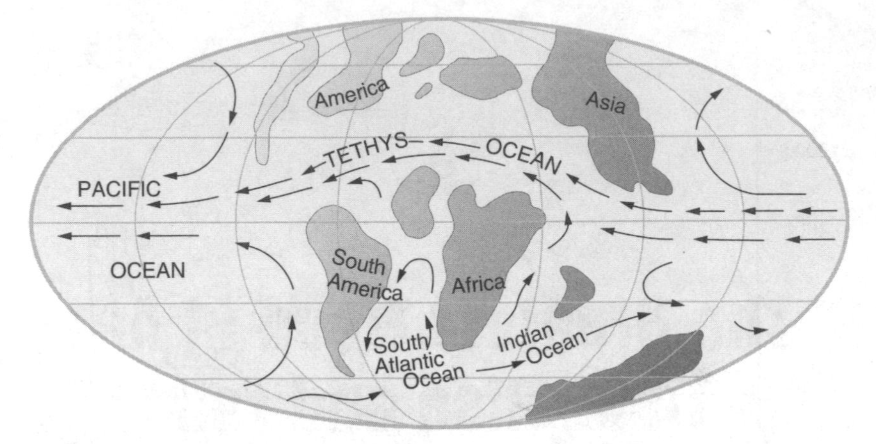

Late Cretaceous Tethys map (80 Mya). Global reconstruction with oceanic circulation, showing the Tethys Ocean at its maximum extent and low-lying continents covering only 18% of the Earth's surface.

An important contributory factor in the Black Death that reigned throughout the Tethys world in the Cretaceous was the rapid growth of the new ocean. The spreading centres from which new ocean floor was formed bulged upwards, forming an impressive mountain chain that was completely submerged beneath the ocean surface. Its enormous bulk, however, displaced ocean waters everywhere, forcing them up and over the adjacent landmasses – sea-level was on the rise, and continued to rise for many millions of years. The sea rose higher than at any point in the past billion years of earth history, perhaps as much as 300 metres higher than today (Fig. 17). The Tethys Ocean, together with her peripheral seas, reached a new zenith, and the whole world was then 82% beneath water. This compares with about 67% today.

Europe was mostly submerged in the late Cretaceous, an arm of the Tethys fingered across northern Africa as the Trans-Saharan Seaway, and the Mowry Sea flooded the North American continent. It was beneath the warm shallow waters of these marginal seas, as

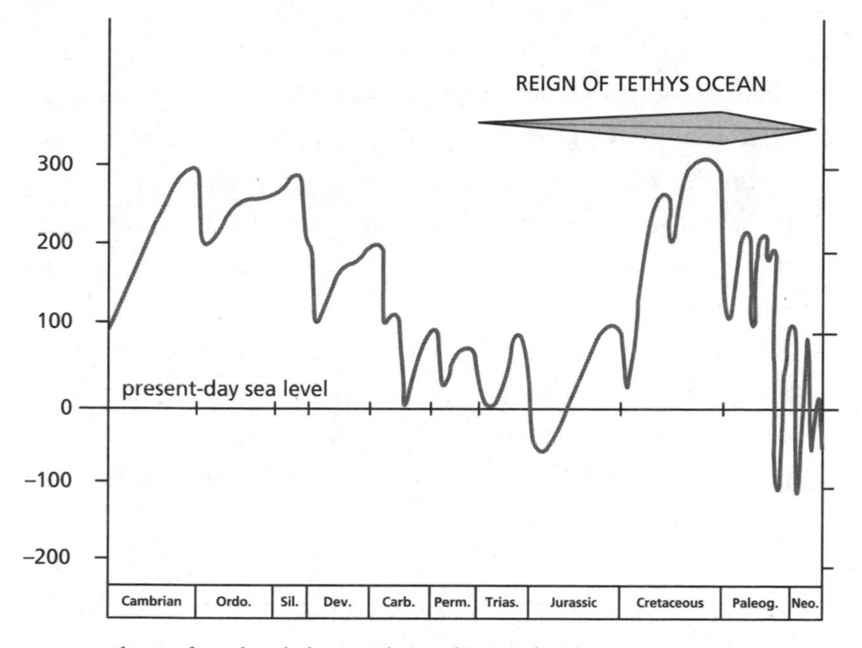

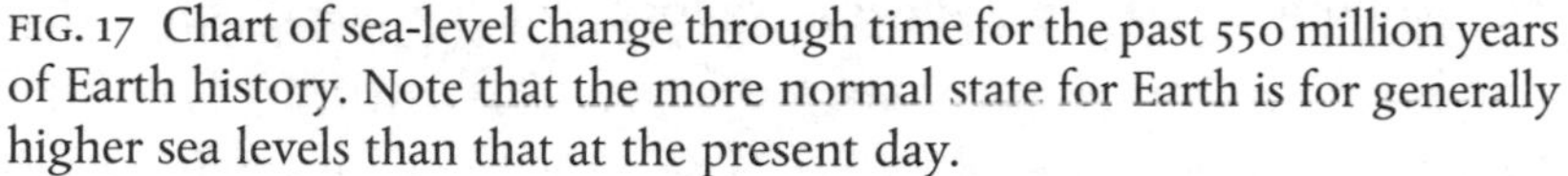
FIG. 17 Chart of sea-level change through time for the past 550 million years of Earth history. Note that the more normal state for Earth is for generally higher sea levels than that at the present day.

well as in the deeper oceans, that soft chalk sediments accumulated, the skeletal debris of a kaleidoscope of microscopic life that blossomed in the plankton. These chalk rocks with their dark bands of flint, now seen everywhere from the Anglo-Paris Basin to North Africa, from Kansas to the Crimean peninsula, display a fundamental rhythm of past climate change linked to the slow ticking of an astronomical clock.

PLATE TECTONICS MATTER

It is time now to revisit the topic of just how oceans grow. Our story began at the end of the Palaeozoic Era, when Pangaea had grown to its maximum size and assumed its single supercontinent status.

This was when the Tethys Ocean first came into being as the large indentation at its eastern margin. But Pangaea underwent an dramatic phase of fracture and rifting, the rifts subsided to such an extent that ocean waters raced in to fill the void, and Tethys cut clean across the heart of a land once dominated by immense mountains and scorching deserts. There were then two great continents separated by a narrow ocean – Laurasia to the north and Gondwana to the south. The narrow ocean grew steadily and widened into the broad Tethys we have reached at this stage of the story, around 80 million years ago. It was at least 4000 kilometres wide, reaching to almost 5000 kilometres at its maximum, equivalent to the width of the North Atlantic Ocean between Great Britain and Canada.

The past distribution of oceans and continents is no mere fanciful tale for dinner-party amusement. Anyone who has read Bill Bryson's delightful romp through time and space, *A Short History of Nearly Everything*, will have been provided with a wealth of information and ammunition to venture an opinion on plate tectonics, as well as on many other topics in science and natural history. Bryson's account of the principal players and events is masterly, from the early dissenters to a dawning realization by the scientific community as a whole of the importance and magnitude of this brave new paradigm in Earth science. In rejecting a paper submitted by Canadian geologist Lawrence Morley in 1963, the editor of the *Journal of Geophysical Research* said: 'Such speculations make interesting talk at cocktail parties, but it is not the sort of thing to be published under serious scientific aegis.' Drummond Matthews and his PhD student Fred Vine, from Cambridge University, came up with more or less the same evidence for the growth of oceans at the same time as Morley, having worked quite independently on magnetic data from the North Atlantic. However, they succeeded in publishing their version in the journal *Nature* in 1963, a seminal paper that has since become accepted as the definitive proof of sea-floor spreading and continental drift.

As an undergraduate student at Cambridge a few years later, I can vividly remember the infectious enthusiasm of Drum Matthews when he lectured to us about these still novel concepts. The buzz of excitement around the Geology Department was almost tangible; several of his younger colleagues, such as Dan Mackenzie and Alan Gilbert-Smith, were themselves to take up the new challenges presented to our science. Plate tectonics is an elegant and simple theory that revolutionized our understanding of the unified earth–ocean system. As students too, we entered into our careers with this new paradigm to the fore, with a new way in which the whole Earth system made sense.

In the preceding chapters, I have indicated just how instrumental plate tectonics has been in Earth's history. The very fusion of plates into a supercontinent was, in part, instrumental in the greatest extinction event of all time. The subsequent break-up and early separation of Pangaea was a significant element in the profligate fecundity of the Jurassic seas. Continued growth of the equatorial Tethys changed ocean circulation, global climate and sea level and led to the Black Death. This in turn has generated the oil wealth on which our frail human society currently depends. There are many more examples. Indeed, in the next chapter it is my contention that plate tectonics and the rumbling interior of Earth are a more likely explanation for the renowned KT extinction event and consequent demise of the dinosaurs, than the more popular theory of extraterrestrial impact. It is time, therefore, that I explained the workings of this unifying paradigm in more detail – at least in so far as new plates are formed and oceans grow.

OCEANIC ASSEMBLY LINE

It was actually a few years before the Vine and Matthews paper that another crucial development in our geological and oceanographic knowledge base occurred, which was instrumental in piecing together plate tectonic theory. This was brought about by the growing use of ever more sophisticated depth sounders at sea. The technological advance was really quite a simple one that uses a loud sound source (a 'pinger'), usually mounted on the ship's hull just below the sea surface, to send a pulse of acoustic energy to the sea-floor. The sound wave bounces off the sediment surface and returns an echo to a receiver adjacent to the pinger. Knowing the speed of sound in water it is a straightforward matter to calculate the water depth from the time taken between the ping and its return. This simple device, whose development was spearheaded by military pressure during the Second World War, as navies and their submarines needed to produce accurate charts of the sea-floor topography, was ably turned to scientific advantage by those privy to such data. One such person was Harry Hess, a Princeton University geologist who had been placed in charge of a US Navy attack transport ship that was fitted with the latest depth-sounding equipment. He later became another of the founding fathers of plate tectonic theory, with his brilliantly perspicacious paper published in the journal *Science* in 1960.

What the new depth sounders had revealed was one of the most remarkable features of the oceans today, almost completely hidden from view some 2.5 kilometres below the ocean surface. This is undoubtedly the world's longest, grandest continuous range of mountains that stretches for some 75,000 kilometres across the face of the planet. From the Arctic Ocean, almost directly beneath the North Pole, a great jagged range runs axially down the length of both the North and South Atlantic Oceans. This range links with another that completely encircles the Antarctic continent, from which great spurs divide into both the Pacific and Indian Oceans. The mountain ranges can be over 1000 kilometres wide and rise over three kilometres above the adjacent ocean floor. This is the global mid-ocean ridge system – the backbone of the oceans, still remote and elusive, untrammelled and unscaled. It has, however, been the subject of scientific scrutiny ever since its discovery.

What we have come to learn from this period of intense investigation, and what was true for the Tethys and all former oceans as well, is that mid-ocean ridges are the sites at which new crust is continually being created. The entire ridge system is essentially a linear belt of active volcanoes and earthquakes within a broad axial rift along which there are very frequent outpourings of a dark volcanic rock called basalt. Hot material wells up beneath the axis of the ridge driven by gigantic convection cells in the mantle – that part of the Earth's interior between its inner core and outer crust. A hot mush of crystals and molten rock, known as magma, accumulates in a large chamber only a few kilometres beneath the sea-floor. Sheets and pipes of this magma force their way through the fractured and weakened crust above and erupt repeatedly into the icy stillness of the deep ocean. Eruption beneath 2.5 kilometres of water is not violent, as it is on land, but more like squeezing toothpaste out of a tube so that the dark, oozing basalt forms a mass of small rounded domes known as pillow lava. The glassy rims of these 'pillows' are testament to their almost instantaneous chilling as lava at over 1000°C meets near-freezing water.

More and more vertical pipes (or dikes) are injected and pillow lava extruded onto the sea-floor, forming the new crust that is gradually torn apart to begin its painfully slow journey across the ocean on a conveyor belt of cooling crustal rock. The elevation of the ridge crest is due to its heat (hence lower density), but as it cools and moves away from the crest it slowly subsides. Older crust, more distant from the ridge, forms the floor of progressively deeper ocean, further weighted down by a steady rain of sediment from surface waters and the influx of material from the continental margin.

The rate of growth globally is slow – some 3.5 square kilometres of new ocean crust is added to the Earth each year – and rates of spreading about the axis are only a few centimetres per year. Over geological time, however, these imperceptible amounts gradually accrue so that whole new oceans are born and grow, ultimately to face destruction in subduction zones as ocean crust slides back down into the mantle through deep narrow trenches in the sea-floor.

Measuring the age of ocean crust yields an intriguing pattern on the sea-floor, and recognition of the systematic increase in age away from the ridge crest provided Vine and Matthews with the final 'proof' of plate tectonic theory in the early 1960s. It is a quite remarkable story of how the Earth's magnetic field interacts with cooling magma, and the sea-floor acts as a magnetic tape recorder. Iron-rich minerals crystallizing out from the hot lava become individually magnetized and oriented towards magnetic north, setting firmly as the lava solidifies. The present direction of the magnetic field is referred to as 'normal' and the opposite direction as 'reverse'. During the geological past, Earth's magnetic field has switched back and forth between normal and reverse with an erratic periodicity, from many thousands to millions of years. Towing sensitive magnetometers behind research ships allows us to measure the sense of magnetization of crustal rocks. The pattern revealed is one of long, narrow bands of normal and reverse magnetic direction with almost perfect symmetry on either side of the ridge crest.

The ages of magnetic reversals have been carefully worked out from radiometric dating of minerals in the magnetized lavas, so it becomes possible to assign ages to the bands of magnetized rocks on the sea-floor. A complete map of the age of ocean crust can be constructed, showing that the oldest crust was formed around 180 million years ago and lies furthest from the present-day spreading centre – the ridge crest. It also shows the direction of plate movement and allows us to calculate the average rate of spreading. In fact, assigning an age to the ocean crust in the deep Angola Basin, by dating the sediments immediately above, was one of our objectives on Leg 75 of the Deep Sea Drilling Program, which I described in the previous chapter while discussing black-shale episodes and the Black Death. On that occasion we determined that the crust we drilled – pillow lava basalts, somewhat altered by interaction with seawater – was around 100 million years old, and that the ocean had begun to spread not long before.

MID-OCEAN ISLANDS

The great majority of any mid-ocean ridge lies deep below the sea surface, such that we had barely any reason even to suspect their existence prior to the detailed bathymetric soundings made in the middle of last century. But we now know that, very rarely, some portions come to rise up higher than all the rest and to peek through the ocean surface as islands above the crest of the ridge. These islands are remote in the extreme and, at the same time, quite enthralling in all that they represent. I can well remember as we steamed on across the South Atlantic from our two months of drilling in the Angola Basin, we slipped past the island of Saint Helena just before dawn. It seemed so desolate, so completely deserted in the pale greyness of almost-light, save for a lone albatross on the wing, as a phantom host of land-starved geologists stared in silence from the port-side gunwales of *Glomar Challenger*. This island was where the British authorities decided to banish Napoleon in 1815, no longer to threaten the rule of empire! He could never have known that St Helena sat astride the mid-ocean ridge, nor what fantastic power rumbled beneath him as he slept.

Iceland is another, much larger portion of the mid-ocean ridge, thrust up and out of the North Atlantic by a long-lived mantle hotspot that has occupied this site since spreading first began. Here it is even possible to walk along a portion of the central axial rift, in Thingvellir Valley, and to see the hardened banks or columns of basalt that have welled up in the spreading centre. Interestingly, Thingvellir was the site of the ancient Icelandic 'parliament'.

I was recently invited to the Azores (April 2009) to participate in a conference on the nature and management of the oceans and to present a paper on the future of energy resources from the deep sea. Now this was an invitation I could not refuse, for the Azores is an island archipelago that sits astride the Mid-Atlantic Ridge some 2500 kilometres due west of Portugal and is the site of another long-lived hotspot. The nine principal islands are rugged, isolated and wholly volcanic. At the western extremity of Faial, which was our conference base, there lies a ruined lighthouse that is now set just inland from the wave-dashed shoreline and draped about with layers of fresh volcanic ash. Some 50 years ago, almost to the day, the azure sea boiled and bubbled as a new volcanic dome heaved upwards and clear above the waves. For month after month the eruption continued intermittently as volcanic rocks and ash spewed outwards from the very bowels of the Earth. A completely new end to the island of Faial was created that year. The sharp cliffs and peak now tower 400 metres above the sea... but all is calm, at least for today.

Climbing those still virgin ash slopes was quite sensational. I imagine that no one before me had ever set foot on that dry scree surface – even the hardy bamboo rhizomes had not yet reached so high, and they seemed to be the first to colonize all else around.

In the alternating light and dark layers of volcanic material of different composition, I could see that some of the blocks heaved out were at least the size of my rucksack – testament to the raw violence that was lurking somewhere deep below. I peered out west from the highest point but couldn't quite make out the rocky islands of Flores and Corvo that I knew to exist some 120 kilometres away. They lie on the other side of the rift valley that runs down the axis of the Mid-Atlantic Ridge – born at the same time as Faial, but spreading with imperceptible slowness in the opposite direction. Flores is often the first port of call for those mad enough to sail single-handed across the Atlantic from America to Europe – my cousin, Adrian, among them. Faial and the other islands to the east of the ridge crest were those visited by Christopher Columbus and his crew on their even more hazardous venture into the unknown, when they crossed the Atlantic 350 years before.

For me, visiting the Azores was especially meaningful as they actually lie on a triple junction in the Earth's great plates, where two arms of the Mid-Atlantic Ridge meet the Azores Fracture Zone. This enormously ancient fracture line runs east across the Atlantic, directly through the Straits of Gibraltar and into the Mediterranean Sea. Here it becomes the gigantic healed suture where the Tethys Ocean finally closed. This suture line was what I initially referred to in Chapter 1, looking out from my window at the Spanish Institute of Oceanography in Fuengirola, where I first began to write this book.

The Tethys Ocean must have grown from a mid-ocean ridge, just like that of today's oceans, expanding until it was the width of the Atlantic Ocean. Fragments of its ocean crust and spreading centre, perhaps even some of the islands that broke the surface of the Tethys Ocean Ridge, have indeed been preserved as ophiolites (Chapter 8).

RISE AND FALL OF SEA LEVEL

At this stage in the history of Tethys, therefore, a huge mid-ocean ridge ran as a jagged submarine mountain belt, somewhere hidden from view along the equatorial region between the continents of Laurasia to the north and Gondwana to the south. This had developed along the axis of one or more of the rifts that originally split Pangaea in two. Mantle hot spots would doubtless have existed, both on and off the ridge itself, perhaps giving rise to islands and archipelagos such as St Helena and the Azores today. To avoid any confusion, I should reiterate here that not all hot spots lie along mid-ocean ridges. Those that were implicated in the rifting and break-up of Pangaea lay beneath a continent; the Hawaiian hot spot in the Pacific Ocean today lies very far from the Mid-Pacific Ridge.

One of the other rifts that originally formed as Pangaea broke apart also began to spread about an incipient ocean ridge, first opening up as a narrow arm of the Tethys, but later evolving into an ocean in its own right – the South Atlantic. Indeed, our drilling on DSDP Leg 75 had shown ocean spreading from just over 100 million years ago and, not long after that, a further burst of volcanic activity (probably hot-spot related) along an ancestral Walvis Ridge. That this gave rise to seamounts and coral-rimmed islands, we know from evidence of their existence in the sediment record we cored. In addition, further to the east, India had rifted from the southern part of Gondwana and what was later to become the Indian Ocean had begun to form.

As I have mentioned before, this period – the late Cretaceous – was a time of extremely high sea level across the world. Although it is difficult to be precise about just how high the sea rose, we are confident that it was at least 200 metres higher than that of today and our best estimates place it at closer to 300 metres higher. In imperial measurements that is 1000 feet of water towering above our heads! About half the area occupied by the continents today would have been submerged – a meagre 18% remaining, most as a low-lying, green and fertile land of islands and small continents. It was these balmy shores against which Tethys' waters lapped. But, before I turn to the marine life that had evolved to populate those seas, I should explain a little more about this whole question of sea level. As we shall see, there is a close link between active sea-floor spreading and the height of sea level globally.

As a young boy growing up in South Devon, I was always fascinated by the tides, especially where there was a broad beach that shoaled gently so that we could body-surf in the waves. Twice daily the sea raced in across the wet rippled sands, and then out again, never once failing – but why did this happen? What forced the tides to come and go, the sea-level to rise and fall? My father explained that it was the result of gravitational pull exerted by the moon and sun on the thin envelope of water we call ocean. And the same tidal forces would also have operated on the waters of the Tethys. Later, as I learnt about the migration of our human ancestors across the southern reaches of the Red Sea and out of Africa for the first time, or across the Bering Strait from Siberia to Alaska and so on to populate the Americas, I was to encounter sea-level changes of altogether greater magnitude and longer timescale. As a budding geologist, I would collect fossils of marine animals from rocks now exposed on the top of cliffs or hills and so came face to face with that age-old problem of how they came to be there – the same problem that drove such antagonistic divisions between early geologists and the Church.

The truth is that, quite apart from diurnal tidal cycles, the shoreline has rarely stood still, but instead has fluctuated widely with the tides of time. Such changes can be inferred from the distribution of terrestrial fossils, such as dinosaurs and plants, in Cretaceous rocks of one region, for example, versus that of marine fossils (marine reptiles or plankton) occurring in rocks of the same age but in a different region. The shoreline must have lain somewhere between the two. In rarer cases, we are lucky enough to find a record of coastal sediments, deposited in deltas, estuaries or beaches, along with fossils of creatures that inhabited such environments (oysters or bird footprints on inter-tidal muds) and hence trace more accurately the position of ancient shorelines. Unfortunately, this sort of evidence is woefully inadequate and often difficult to access where it is underwater beneath the shelf seas of today.

However, thanks to a further development of the depth sounding method, as used to determine the depth of the sea-floor and shape of its underwater relief, we are now able to interrogate the thick piles of sediment beneath continental margins, which had accumulated slowly and continuously as the oceans widened and their margins subsided. By using a more powerful source of energy with a lower frequency, part of the energy will penetrate through the sea-floor and be reflected from boundaries between the different layers of sediment. This technique – since much refined by the oil industry – produces fantastic visual records of the subsurface, known as seismic profiles. These are like geological cross sections showing the succession and arrangement of different sedimentary layers, from which it is possible to extract the changing position of the coastline through time – and hence the rise and fall of sea level. The record at sea is certainly much more complete than that from scattered outcrops on land and, although the details of such records are hotly disputed by geologists, the general pattern is accepted.

I will concern myself mainly with what has happened to sea level during the existence of the Tethys Ocean – at least for the 250 million years of its history that I am charting in this book. During the late Permian period, sea level was dropping to one of its all-time lowest points in the past half billion years of the Phanerozoic Era. This was coincident with final construction of a single supercontinent, as well as with the great Permian extinction. Over the ensuing 170 to 180 million years, sea level rose by over 300 metres to its maximum level in the late Cretaceous, in a series of 10 to 12 stepped increases and smaller decreases known as supercycles. Since that time, sea level has fallen through a similar series of supercycles to reach its present relatively low level. The whole of the Tethys, therefore, existed through this one megacycle of change. When looked at in still more detail, the supercycles are superimposed on cycles of shorter time-scale and these, in parts, on still higher order cycles of change (Fig. 17).

The forces that drive these long- and short-term cyclic changes in sea level are not simple to explain. We believe we have part of the story as cogently argued by Cambridge geologist, Professor Tjeerd van Andel, in his delightful and informative book *New Views On An Old Planet*. But Tjeerd goes on to say of the search for an explanation: 'Not only are the questions far from being answered, there is not even unanimity about which are the right questions.' I hasten to add that this in no way casts aspersion on that part of the causal mechanism we do understand.

There are three principal causes of global sea-level change of which we can be confident. The first of these is directly related to sea-floor spreading and, most particularly, to the total amount and rate of spreading. Rapid rates of spreading create larger volumes of new hot oceanic crust than do slower rates. This hotter crust is less dense and so rises up to displace a greater volume of seawater such that sea level rises. Exactly the same effect is produced by extending the overall length of mid-ocean ridges in existence at any one time. This scenario explains the first half of the Tethys megacycle of sea-level rise. New oceans opened up, rates of spreading were rapid, and the overall length of the mid-ocean ridges increased. A corollary was the crumpling together of continents and pushing up of mountain ranges that occurred when Pangaea fused together, so that there was even less continental mass in the oceans, thus further contributing to lowered sea level at the start of the megacycle.

The second causal factor is also linked to plate tectonics, specifically to the nature of the ocean–continent transition. There are a range of parameters involved, including how far away the spreading centre lies, whether or not subduction has begun, if a deep-sea trench has been created, and how far from the coast is there a high mountain range. It is rather more complex to work through the overlapping effects of these different parameters but suffice it to say that this combination of factors could explain the occurrence of variable duration supercycles.

The third factor is the one with which most people are more familiar – that is the locking up of large volumes of seawater as icecaps on land. Note that, for this to have any effect on sea level, the ice must be locked up on the continent, rather than simply floating at sea, which would create a neutral effect. Certainly we can document very accurately the fluctuation in sea level over the past two million years of glacial–interglacial conditions. Dramatic though they have been, particularly for the survival and evolution of our hominid ancestors, these are but high amplitude, short-term fluctuations that have characterized the geologically recent past. The Tethys Ocean was long gone by this time. Indeed, the world was more or less ice-free for the whole time that Tethys existed. Therefore, in order to explain the second half of the Tethyan megacycle, that of gradual sea-level fall since the Cretaceous acme, we must return to plate-tectonic explanations, reduction in rates of spreading and eventual closure of the Tethys.

In addition to sea-level fluctuation that is truly global in its extent, the picture is further complicated by local changes, which may affect only one small area or coastline. These are generally more readily explained as the result of tectonic uplift and subsidence (i.e. local earth movements) and by the sudden deluge of sediment eroded from newly uplifted mountain ranges and deposited in large deltas building outwards from the coast. The loading and unloading of a region by ice build-up and then decline, first serves to depress the land locally and then allows it to rebound after the icecap has gone.

THE FLOOD ENCROACHES

Just as the continental red-beds of desert and river origin were so widespread across the Pangaean supercontinent during Permian and Triassic time, so are the white chalk cliffs of almost every continent immediately recognizable as being of the Cretaceous period. Indeed, the name itself comes from the Latin word '*creta*' meaning chalk – a type of very fine-grained, pure limestone, which is often rather soft and crumbly on the exterior surface.

Travelling across many parts of the world today, I have encountered this same rock type as clear evidence of the dramatic level to which the late Cretaceous Tethys eventually flooded – truly a leviathan among oceans. I have always found it to be slightly perplexing and unique, both as a rock and in the landscape it creates. Chalk hills are at once gentle and rolling, but with a scruffy soil and hence harsh, sparse vegetation. Cliff lines are steep, sharp and dramatic, but the rock itself crumbles softly to the touch. In blistering sunshine, it remains cool to the touch, is one of the only rocks I have to observe with dark glasses, and is often speckled with irregular black flint nodules. Rivers or beaches today can abound with these half-rounded flint pebbles, but the chalk uplands from which they derived may be long gone – dissolved and swept out to sea.

As an intriguing aside, I have often wondered how far the human ape would have developed in the use and refinement of stone tools had it not been for the readily available flints so intimately associated with Tethyan chalk. Or, for that matter, how we would have gone on to create champagne and the subtle Sauvignon blanc wines (my favourite) without the chalk uplands of the Paris Basin.

Now let me be a little more specific in just what these chalk hills can tell us of the Tethys Ocean, besides her obviously broad extent. The axis of the Mediterranean lands – Crete, aptly named by the Romans, Corsica, Calabria, Sicily and several of the Greek islands – all boast chalk from the central Tethys. It is thick, very fine-grained and remarkably monotonous. Even with the conventional petrographic (rock) microscope, perhaps the next most invaluable tool for a geologist after a hammer and hand lens, it is generally impossible to see anything at all. I was still a student at Cambridge when the new technique of scanning electron microscopy was just becoming available, at least to the richer and better-equipped universities. One of our lecturers at the time was working on the nature of chalk and showed us fantastic new images of the tiny one micron diameter coccolith platelets of which it is mostly composed. That is one-thousandth of a millimetre – insanely minuscule, and yet each layer of chalk across the Tethys is composed of countless trillions of these white platelets of calcium carbonate (Fig. 18A).

Travelling east through the Carpathians to the Black Sea and beyond, or south through the Middle East, and across North Africa – chalk is everywhere. So too did it spread over great tracts of what had hitherto been continental landmass, and which had become relatively flat and low-lying after years of persistent erosion. From Libya and Tunisia, Tethys swept clean across the African continent to Nigeria and spilled into the rapidly growing South Atlantic. From the Gulf of Mexico, it spilled north through Texas, Alabama and Colorado to link with the Mowry Sea further north and hence opened into the Arctic Ocean. Interestingly, the modern

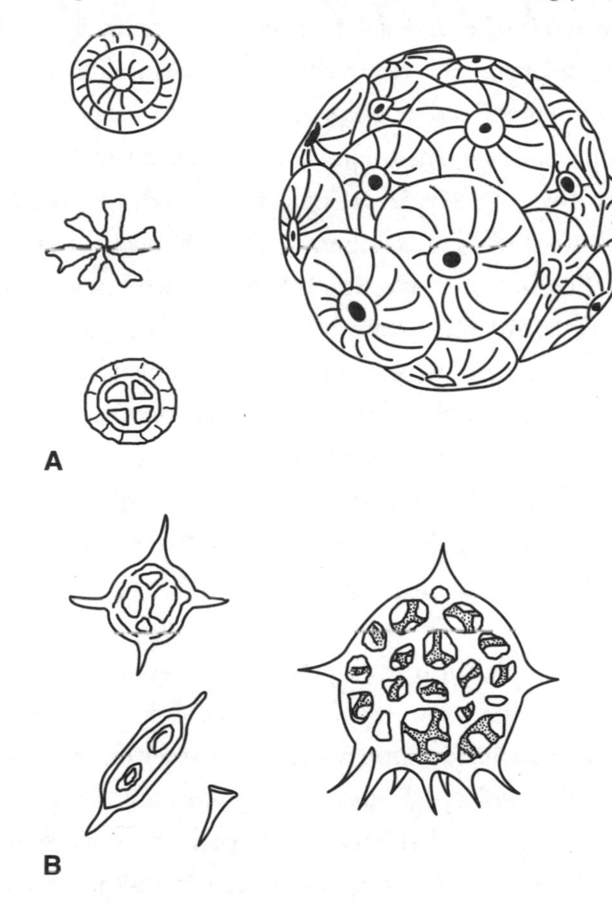

FIG. 18 Drawings of Cretaceous microfossils from the Tethyan chalk succession. A – a spherical coccolithophore made up of overlapping calcareous discs or plates (coccoliths); some of the different plate types shown on left. Scale bar = 5 microns (0.005 mm). These, together with foraminifers (Fig. 14), are the main component of chalk. B – three types of silicoflagellate made of clear silica glass. Scale bar = 30 microns (0.03 mm). These, together with radiolarians and diatoms (Fig.14), are the main component of flint nodules and bands.

style of tropical coral reef that had taken over the Jurassic seas was replaced in the late Cretaceous by rudist bivalves. These were an unusual type of mollusc with a cone-shaped lower shell and lid-like upper shell; they grew together in very large numbers, and seemed to have defeated the hexacorals temporarily in the competition for space. At higher latitudes, Tethys' chalk and rudist reefs gave way to sandier shorelines and Arctic clays on the sea-floor.

One of the classic regions of northern Europe covered with these chalk seas is the Anglo-Paris Basin. On both sides of the English Channel (or *La Manche*, according to the French) there are imposing white cliffs: the Cliffs of Dover, Seven Sisters of Kent, and *Les Falaises d'Etretat* in northern France are all chalk. They form the North and South Downs of southern England, with their crystal clear streams and quintessentially English villages. They disappear beneath the North Sea only to reappear in Denmark. Nowhere are they very different from those throughout the Tethys, though perhaps even more distinctive for the number and persistence of black flints that occur in more or less regularly spaced bands.

CHALK – FLINT CYCLES

These flints actually have an interesting and rather different story of their own. Both coccoliths, with their tiny skeletons of calcite, and diatoms, which constructed their shells of clear silica glass, had become absolutely dominant in the ocean plankton. They were joined by armies of voracious zooplankton – foraminifers, calpionellids and silicoflagellates (Fig. 18B). The constant rain of this mixed skeletal debris covered the sea-floor in soft white ooze. However, the type of silica used by diatoms and silicoflagellates for their ornate living carapaces is opal, which soon becomes chemically unstable when separated from living tissue and buried in sediment beneath the sea-floor. The opaline silica dissolves into pore waters within the sediment and then later precipitates out, when the chemical conditions have changed sufficiently, centring around a nucleus to form irregular-shaped flint nodules comprising only the ultrastable form of silica known as quartz. Tiny impurities in the quartz crystal lattice give the flint its distinctive colour – traces of organic matter yield black flint, whereas traces of iron give brown and yellow hues.

Even the regular alternation of chalk–flint bands tells us something of the stability and climate in those halcyon seas (Fig. 19). The banding was most likely due to an original periodicity in the flux of diatoms to the sea-floor, which in turn, was driven by long-term climate cycles related to periodic changes in the way the Earth circles the sun. It was a Serbian mathematician, Milan Milankovitch, who, as we

FIG. 19 Photograph of Chalk cliffs on the Isle of Wight, southern England, showing distinct climate-controlled bedding and black flint nodules (in foreground) (Photo by Claire Ashford).

have seen, demonstrated subtle changes in three different aspects of Earth's orbit: (1) the degree of eccentricity in its elliptical orbit varies by just 6% over a period of 100,000 years; (2) the tilt of Earth's axis varies by just a few degrees over a period of 41,000 years; and (3) Earth actually wobbles rather like a top as it moves around the sun, but very slowly with a period of 22,000 years. The rather complicated interaction between these different cycles results in very slight but significant differences in global temperature. This scale of cycles, now known as Milankovitch cyclicity, was first used to explain glacial–interglacial periods of the last ice age. However, it is now implicated in much more besides, in other amazing tales from the Tethys and elsewhere that I must leave for now.

There are two further fascinations I have with flint. The first is the constant allure of finding a Palaeolithic stone tool in a hidden corner of the world where none has been found before. The second is the way the quirky Englishman, from Saxon times and before, has grappled with such completely unworkable and irregular a building stone and yet succeeded in constructing the most delightful and durable churches, grandiose houses and lengthy stone walls.

OLD AND NEW IN A GREENHOUSE WORLD

So this was Damascus, the oldest continuously inhabited city in the world. I love the imposing hills of hard fractured Cretaceous limestone that dominate the city rim, and against which the blocks of flats become progressively smaller, older and shabbier, distinctly more troubled by the passage of time and lack of money, as they rise up the beige–white flanks of chalk. In the distance, it is hard to distinguish rock from residence. In front of me, and within these walls of time, the rich pastiche of history was breathtaking – narrow streets and noisy souks, a majestic line of Roman columns that end suddenly by propping up a corner tea shop with its pitifully shabby chairs and somewhat unwelcome smells from the night before, pavements that are cracked and bustling – in such contrast with one of the most tranquil and beautiful mosques I have ever seen.

What brought me to Syria in 2009 is pertinent to the Tethys story in two respects. The first was to see in the field some recent and exciting finds of late Cretaceous marine vertebrates in the Palmyrides Mountains. The second was to work along the Euphrates River on fluvial rocks of a much later age, swept down from the mountains that formed at the final closure of Tethys, but which I shall leave aside for now. Working with a small team from Al Furat Petroleum Company (a Shell-Syria joint venture) and the University of Damascus, we joined the bus chartered daily by Al Furat in order to ferry workers to and from their operations in the eastern provinces. The road signs within the city show Lebanon and Beirut to the west, Jordan a little further south, and even a small sign pointing north to Turkey. Straight ahead to the east lies the border with Iraq – not that we were to travel quite that far, for the first plan was to alight at Palmyra.

The road to the east appears to run forever, through little more than barren stony desert. Endless and straight, save for a few gentle bends that would seem designed only to break the monotony for truckers and bus drivers on their long lonely treks across such a bleak unforgiving land. Certainly it seemed to me that the road was a shade too narrow for the thundering onslaught of trucks heading towards Damascus as we hurtled in the opposite direction!

Then, quite suddenly, we came upon Palmyra – the quiet rows of standing columns, each with magnificent Corinthian capitals sporting such ornate detail, even a double colonnade with imposing archway, an amphitheatre, temples, spacious courtyards . . . all have stood the test of time, part protected by a dry desert climate, part ravaged by the peoples who have dwelt here through long annals of history – but they have survived. There are fallen remnants too, strewn everywhere across the land, each with its own particular magic, its own tale to tell. There can be little doubt that this was once a magnificent city indeed, especially in its Roman heyday between the second and third centuries, when it rose to great prominence as the halfway caravan city on the favoured, safer and shorter, desert trade route between Doura Europa on the Euphrates, gateway to eastern treasures, and Antioch on the Mediterranean. But it is also one of the larger desert oases, with plentiful subterranean springs, and has been continuously inhabited for over three millennia. We were truly in a timeless land.

From there we changed transport for a short trek into the Palmyrides chain, heading towards the phosphate mines at Charquieh and Khneifiss, and on to further outcrops in the chalk hills near Bardeh and Soukhneh. The deposits were rather scruffy and very dusty, as quarries and chalk generally are, but the finds quite outstanding. More than 50 different species of marine vertebrates have been recovered, which together provide an excellent snapshot of life in the central gyres of Tethys – and show a rather curious mixture of modern and archaic forms. There were large numbers of sharks, rays and other teleost fishes (i.e. essentially modern types) sharing the same top-predator niche with reptilian sea monsters, including plesiosaurs and mosasaurs that would shortly become extinct. These last were huge marine lizards that grew to over 15 metres in length and possessed rows of ferocious pointed teeth. (Note that these were quite different from the much smaller and elegant mesosaurs that I found in the Parana Basin of Brazil and which died out in the great Permian extinction.) The mosasaurs appear to have flourished towards the end of the Cretaceous period while most of their reptilian cousins were in decline. They were joined in the spoil of fossils by some of the more ancient crocodiles and a host of modern marine turtles, on the scene for the first time in the late Cretaceous, some growing to a length of nearly 4 metres.

This was clearly not a deep-water assemblage and yet palaeogeographic reconstruction suggests that Syria lay almost 2500 kilometres from the nearest landfall, which would have been somewhere on the central African continent. The phosphatic sediments within which the most abundant finds were made are typical of outer shelf conditions, with extremely high primary productivity and generally low oxygen levels. Clearly, there were very broad continental shelves extending far into the ocean basin.

These were golden days in a truly greenhouse world. Both the average temperature and the mean sea level were at an extreme high across the planet. The ocean surface was a cornucopia of plankton, probably the most productive the world has ever seen. Perhaps it was all too good to last.

7

End of an Era: The Debate Continues

Rogue of the high seas
I have seen where you touched
These lava islands
Blackened, rugged
Piercing ruffled seas
Witness at the birth of Earth's
Giant tectonic plates
Waves that beat and break
Eroding shores of a new world

From *Adrian* by Dorrik Stow

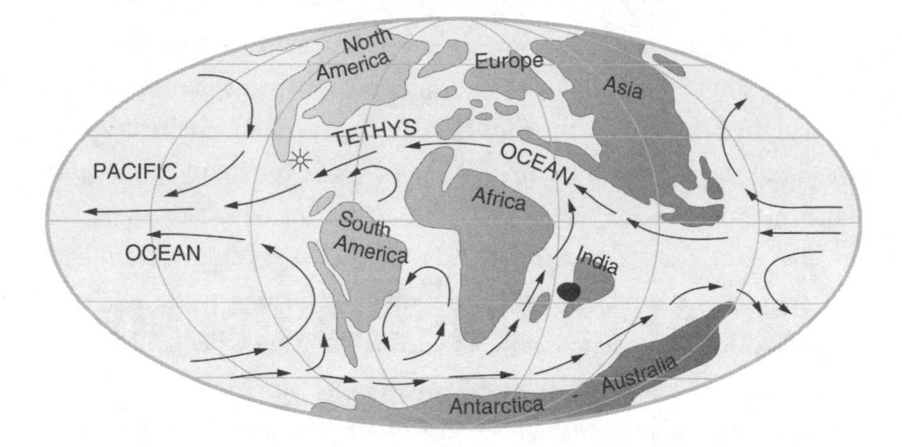

End Cretaceous Tethys map (65 Mya). Global reconstruction with oceanic circulation, showing the Tethys Ocean at the KT extinction event. Note the position of the Chixuclub impact crater and the Deccan superplume volcanic event.

The end of the Mesozoic Era came abruptly but was not altogether unexpected. It is perhaps the most widely known and publicly debated moment in the whole history of geological time, being marked by the mass extinction event in which the last of the dinosaurs disappeared and so finally brought to a close the great Age of Reptiles. For about 180 million years dinosaurs and their close relatives had ruled the land, terrorized the seas and fought for prowess in the sky. But, by the end of the Cretaceous period, some 65 million years ago, their time had come. With them went most of the marsupial mammals and birds, and roughly a quarter of all species of crocodiles, turtles and fish. Altogether about 20% of known families and an estimated 50% of all species became extinct. The duration of this KT boundary event is known to be relatively short, but was not instantaneous. Although our understanding is hampered by the relative paucity of good sections on land through which we can piece together an accurate picture of animal life and death, it suffers even more from the blinkered polarization of views that characterizes most debates on the subject.

Literally hundreds of theories have been proposed for the death of the dinosaurs – some with an element of sense, but many more quite fanciful. Suggestions that their brains were too small to meet the challenge of change, that mammals ate their eggs, that caterpillars ate their plant food, that they became overheated, suffered from slipped discs, toothache or eye infections, that they became constipated or were killed off by poisonous fungi – are all fun but fail to consider why many other groups of plants and animals died out at the same time. One that I chanced upon recently has to be among my favourites for implausibility – it suggests that dinosaurs became so hot, bothered and cramped for space on the few islands remaining amid such wide expanses of sea that they simply refused to copulate! There are countless more, but as none are based on recourse to the accumulation of facts, they are not science and so need not concern us here.

Although the fate of dinosaurs has understandably caught the public eye, by far the most significant extinction at this time was that of the myriad coccoliths and related planktonic species that populated the open ocean – including Tethys, and all her medusoid arms. They proliferated in such unimaginable abundance that their remains are seen everywhere to this day as chalk cliffs, uplands and even high in mountain crags. The life and times of these abundant seas was the principal subject of the previous chapter. But it is not just because of their very large numbers that I emphasize their significance, but rather due to their prime position at the base of a rich food chain – truly the foodstuff of the world.

The aim of this chapter is to take a dispassionate and scientific look at what is really known about the mass extinction that characterizes the end of the Mesozoic Era. Remember, this is also the end of the Cretaceous period and is dated at 65 Mya. As I have mentioned before, it is also commonly referred to as the KT extinction or KT boundary event, using a Germanic-English acronym of the words *Kreide* (meaning Cretaceous) and *Tertiary*, for the era that followed.

WHERE HAD ALL THE LAND GONE?

Let me recap very briefly, from Chapter 7, what we know about the land areas at this time. They covered no more than 18% of the Earth's surface area – a little more than half the area of continent today – and were, for the most part, low-lying and peneplained (much flattened by long-term erosion). The principal mountain ranges lay along the western coasts of North and South America, much as the Rockies and Andes today, and along the eastern margins of South East Asia and north-west Australia. For, while Tethys had been spreading, the Panthalassic Ocean (or Pacific Ocean) was undergoing subduction beneath the outer rim of the continents. (It is worth mentioning here that most palaeogeographic reconstructions, including those I have placed at the beginning of each chapter, drop the use of Panthalassa either after the Triassic or Jurassic Periods, and replace it by the Pacific Ocean.) There was no landmass over either of the polar regions. Equally significant, the small continental areas that did exist, having steadily drifted apart through the Cretaceous period, were geographically well separated. North America was divided into two or three segments, and separate from South America. Africa was divided into two, India was on its own in the middle of the Tethys, while most of Europe, south and central Asia were submerged. Siberia was joined together with South East Asia, and Australia was still joined with Antarctica; these were the largest continents of the time. Australia was unique among the land areas, in that it did not become submerged when extremely high sea levels caused maximum flooding elsewhere. Some tectonic peculiarity we do not yet understand was pushing the whole land area of Australia upwards.

Having established where the land areas were located, let us consider their inhabitants. What this geographic separation meant was that isolation and biozonation of land-based organisms occurred rapidly. This was equally true of coastal and shallow marine groups where separation between landmasses was sufficiently wide and deep to preclude migration. Certainly the dinosaurs and pterosaurs, which were already showing signs of environmental stress, were affected in this way. Those groups that had been living in Africa, Asia and South America during the early Cretaceous were in decline, and none survived into the late Cretaceous. The final rump of survivors was confined to western North America, boasting high diversity some 10 million years before the end, but then reducing to a mere handful of species as the KT boundary was approached. In fact, the ultimate demise of these magnificent and fearsome beasts was more of a whimper than a bang, which does not augur well for a catastrophist explanation.

Less is really known about the rate of decline of most other terrestrial species, largely owing to the lack of suitable places and conditions for fossil preservation, so that even simple statistics on what finds have been made become fairly meaningless. We estimate that three-quarters of birds and marsupial mammals became extinct, along with a quarter of the large crocodiles and turtles. But lizards, snakes, amphibians and the great majority of placental mammals remained more or less unscathed. How could this have been so?

Before leaving the land and returning to the 'comfort' of the seas, I must mention one fundamental change that had taken place, at first along the edges of the continents and close to the river banks, but soon to spread far and wide and to alter the face of the world forever.

This was the evolution of flowering plants – or angiosperms, meaning 'encased seeds' rather than the 'naked seeds' of gymnosperms. It was a mild revolution without fanfare, which took place some time during the mid-Cretaceous on the fringe of continents already clothed in rich greenery of ferns and conifers, while the Black Death wreaked havoc in the seas. There were no very recognizable flowers to begin with and, after a while, there were probably white flowers. The rich palette of colours we associate with a spring meadow, or the extravagant largesse of a tropical rainforest, came later. This expansion into colour was driven by the co-evolution of angiosperms with animals, most importantly the world of insects. But once the revolution had taken hold, the adaptive radiation and diversification of flowering plants on those rumps of continent remaining above sea level occurred with startling alacrity.

All along the Tethys shoreline of the North American landmass at this time were deposited continental sands, silts and variegated muds full of the black carbonaceous remains of early flowering plants. In a well-hidden series, which today outcrops beneath the Atlantic coastal plain of Maryland not far from the busy city life of Washington and Baltimore, there is a treasure trove of these embryonic angiosperm fossils. Through a 10-million-year time slice, the diversification of new pollen and leaf types is preserved in immaculate detail. Between that time and the KT event, angiosperms went on to surpass gymnosperms in their variety and adaptability. As many as 50 complete families survived the end Cretaceous unscathed, including the sycamore and plane trees, hollies and magnolias, oaks and walnuts, birches and alders. Forests of the late Cretaceous were already beginning to look relatively familiar. In some ways, it seemed as though this radiation was almost instantaneous on a geological timescale – like a mass extinction in reverse – leading Charles Darwin to consider it 'an abominable mystery'. The open, unforested areas, on the other hand, would have looked quite strange for the grasses had not yet evolved, although primitive forms of heathers, mallows, myrtles, spurges and nettles were already a firm part of this nascent and richer tapestry.

Why were these newcomers so successful and what led to this particular evolutionary development? The first of these questions is easier to answer. In angiosperms, the seed is fertilized within the encased ovary, and the surrounding fleshy fruit provides the seed with a ready supply of nutrition. This enabled a much more rapid reproductive cycle than for gymnosperms, a secure start in life for the germinated seed and hence the ability for opportunistic colonization of a changing landscape. Both the fruit and the development of flowers enlisted insects, birds and other animals to aid in the wider spread of pollen as well as its targeted dispersal to like flowers living in a different area. The rewards of sugary nectar and succulent fruit, easy to produce as a by-product of photosynthesis, were a small price to pay for such advantage over their competitors. Edible nuts and seeds were another development in this fascinating chapter of evolution played out along the shorelines of Tethys. The fact that these specific traits of angiosperms were so immediately successful suggests at least a general answer to the second question: the global environment was changing and therefore stressful for the established plants and animals of the time.

STRESS AMONG THE SEAGRASS

As an indication of their supreme adaptability, one member of the angiosperms, the seagrasses, even managed to colonize successfully the marine environment during the late Cretaceous. These grass-like plants carpeted the shallow sea-floor, learning both to live in salt-water and to find an effective means of subaqueous pollination and seed dispersal. They survive today as eel grass, widely distributed throughout the North Atlantic and Pacific, and as turtle grass and manatee grass, which form dense underwater meadows throughout the Caribbean. They provide a unique and essential habitat for a host of benthic organisms, crustaceans and fishes, as well as being the primary food for green turtles, dugongs and manatees.

These last, the dugongs and manatees, are both vegetarian marine mammals of the order *Sirenia*. The oldest fossil sirenians found so far date from around 50 million years ago, following a similar evolutionary trend to the cetaceans (whales and dolphins) – a theme I shall return to in Chapter 9. Incidentally, the 'sirens' in ancient mythology were sea nymphs or mermaids who lured sailors and their ships to treacherous rocks with their seductive appearance and mesmerizing songs. With much imagination, no doubt fuelled by long months at sea, the broad barrel-shaped manatees swimming among seagrasses may have borne some resemblance to the human form and so helped to perpetuate the myth of mermaids.

Although sirenians had not yet evolved in the late Cretaceous seas, turtles certainly had, including the giant *Archelon*, a probable forerunner of the green turtle today. My palaeontology colleague from Southampton University, Dr Ian Harding, informs me that very modern-looking crabs also evolved at this time, although whether their bizarre sideways gait and oversized waving pincers could ever look 'modern', I would dispute. Many new types of bivalve and gastropod found safe haven among the waving seagrasses and the mangrove roots, or in cavities and crevices of nearby reefs. Fossil pollen from the mangrove palm *Nypa* has been found in late Cretaceous rocks of North Africa, indicating that mangroves had evolved to join the seagrasses, at least along the southern shores of the Tethys. While some of the small, brightly coloured sea snails grazed happily on coralline algae or seagrass fronds, others were playing out their thoroughly gruesome carnivorous lives, attacking and externally digesting much larger prey than themselves.

Another larger and more elegant predator evolved in the late Cretaceous to take advantage of the bounty among the seagrasses. This was the flightless diving bird *Hesperornis*, of which the near complete fossilized skeleton has been found. The diminutive wings were no longer of use in flight, but had been adapted for swimming, aided by large paddle-like feet. Its sharp, backward-directed teeth are a clear sign that it had a predilection for small slippery fish.

But it was not all seagrasses. Rudist-bryozoan reefs were new on the scene and, for a short spell of late-Cretaceous time, completely usurped coral and algal reefs, a successful combination that had flourished through the previous 95% of the Mesozoic Era. The rudists were simple bivalves with tough, greatly thickened, cone-shaped lower valves. They stuck together (quite literally) in great numbers, firmly cementing themselves to each other and to some other giant bivalves of the oyster family, which had also made their presence felt in this final flourish of the Cretaceous. They certainly make unusual and attractive building stones, and have done so since Roman times. While working in Sicily and then on the Greek island of Cephalonia, I chanced upon ancient, long-since-abandoned quarries in rudist limestone, almost certainly of Roman origin, perhaps even Greek on Cephalonia. On both occasions, the students with me were at first very puzzled by a limestone with a distinct framework that was not coralline.

The success of the rudists and their companions was doubtless due to an effective partnership with symbiotic algae or another micro-organism that lived and multiplied in their tissues. Although direct fossil evidence for such a relationship is most unlikely to survive, we can surmise as much from the biology of the corals, giant clams and other bivalves that build up the foundations of today's reef communities. For as well as being principal reef-building organisms, they are also the dominant primary producers at the base of these richly complex food webs. This dual role is only possible because the tiny individual coral polyps are symbiotic hosts to a single-celled dinoflagellate species known as zooxanthellae. It is these microscopic organisms that are the photosynthesizing member of a very close partnership, in which they line the inner wall of the polyp tissue. Each measures about 0.01 mm across so that there are as many as 1 million cells in just a square centimetre of coral. The zooxanthellae receive a safe home in a warm light environment, further protected by waving rows of stinging cells, while the coral animals can grow and flourish with an ever-ready supply of internal nourishment. Each polyp is further able to extract calcium carbonate from seawater and to secrete this as a hard lime cement cup surrounding its body.

A significant problem for these magnificent ecosystems today is that of coral bleaching. In perfectly healthy reef complexes a thin living tissue of coral polyp and zooxanthellae stretches over the limy skeleton of the entire colony, frequently coloured pink, red, green, yellow, brown or purple. However, at some increased level of environmental stress – probably higher sea-surface temperatures caused by global warming – the coral polyps expel their zooxanthellae symbionts and both die. The result is bleached white coral that may be beautiful to collect but is quite lifeless.

Was this what happened in the late Cretaceous Tethys, when corals were replaced by rudists? Did the same happen again with rudist reefs and their new microbial symbionts? I would suggest that the answer was in the affirmative for both cases. Today's coral megacities cover just 0.2% of the ocean floor but support around 25% of all known marine species; we assume that it must have been similar for the Tethys coral and rudist reefs. Removing the very fabric of life from these cornucopias would have had a very significant knock-on effect on the whole complex community. Just how the evident environmental stress (or stresses) at the time affected the seagrass ecosystems is less clear from the fossil record, largely because the seagrass itself is not easily fossilized.

However, what we can ascertain is that most coral species died out between 10 and 20 million years before the KT boundary. While rudists proliferated to fill the gap, it was only one or two sole species that survived until the end, for they too showed a very marked decline over at least the last 2 million years of the Cretaceous. The gastropods that had come to flourish in the warm Tethys waters, amid the seagrasses and reefs, suffered badly. All those species from across the North African region died out and were replaced by cooler-water species that had migrated south from Greenland. Most brachiopods, adapted for life in warm chalky seas of Europe, showed very rapid decline in the final years of the Cretaceous Period. They too did not like the changes that were afoot (Fig. 20) – the potential causal factors of which we shall return to shortly.

ALARM IN THE OPEN OCEAN

Among the larger, free-swimming creatures that both decorated and terrorized the Tethys Ocean, marine reptiles, coiled ammonites and squid-like belemnites epitomize the Mesozoic Era. How then were these creatures faring towards the end of the Cretaceous Period? Despite some remarkable finds of marine reptiles, such as those in the phosphorite quarries of Syria that I discussed in the previous chapter, the fossil record is too sparse to make any confident assertions about this whole group. It does appear that many reptiles, such as the ichthyosaurs of Jurassic fame, had already disappeared well

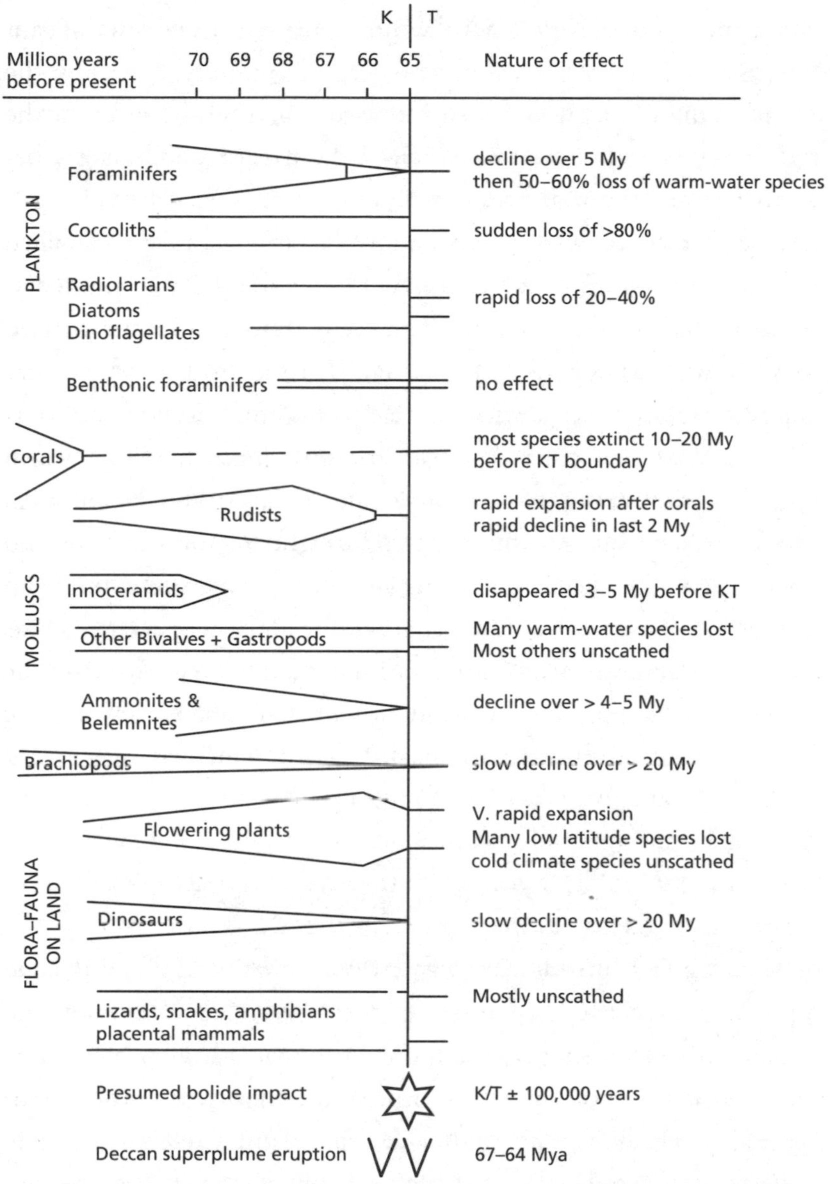

K T

Million years before present 70 69 68 67 66 65 Nature of effect

PLANKTON

Foraminifers — decline over 5 My then 50–60% loss of warm-water species

Coccoliths — sudden loss of >80%

Radiolarians
Diatoms
Dinoflagellates — rapid loss of 20–40%

Benthonic foraminifers — no effect

Corals — most species extinct 10–20 My before KT boundary

Rudists — rapid expansion after corals rapid decline in last 2 My

MOLLUSCS

Innoceramids — disappeared 3–5 My before KT

Other Bivalves + Gastropods — Many warm-water species lost Most others unscathed

Ammonites & Belemnites — decline over > 4–5 My

Brachiopods — slow decline over > 20 My

FLORA–FAUNA ON LAND

Flowering plants — V. rapid expansion Many low latitude species lost cold climate species unscathed

Dinosaurs — slow decline over > 20 My

Lizards, snakes, amphibians placental mammals — Mostly unscathed

Presumed bolide impact — K/T ± 100,000 years

Deccan superplume eruption — 67–64 Mya

FIG. 20 Chart showing the gradual nature of extinctions for a number of different animal and plant groups leading up to the KT boundary.

before the end of the Era, while the mosasaurs flourished briefly to fill the gap.

I have hunted high and low, but have only ever found scrappy remains and disarticulated vertebrae of ichthyosaur fossils. However, I have been far more successful with ammonites (Fig. 21) and belemnites (ammonoids), which are far more common as fossils throughout the Mesozoic. They were particularly prevalent during the acme of late Cretaceous flooding, and their record tells us something more of what was happening at this time. There is a particularly well-exposed sequence of deep-sea sediments with ammonite fossils along the coast of northern Spain, near Zumaya. This is a truly hidden corner of sumptuous seafood and fine Spanish wine – my own favourite being the Albarino whites brought in from the Galician region just along the coast. Even more exciting, the sediments record a cyclic alternation of turbidites and pelagites that span across the KT boundary. These are sediments reflecting different depositional processes in the deep sea – turbidites are instantaneous flow events, whereas pelagites represent the normal background rain of material to the sea-floor. Fossils are best found in the latter. There can be no doubt that the ammonoids show a distinct pattern of gradual decline over some 4 to 5 million years, with the very last fossil found about 12 metres below the top of the Cretaceous strata. From the average rate at which these sediments accumulated, we can estimate that this latest ammonite lived and died 100,000 years before the Mesozoic Era ended. This series simply exemplifies the story that emerges elsewhere in the world. Only 20 ammonoid species are known from the last two million years before the end of the Cretaceous; about twice that number lived during the two million years before that.

And curious changes were also taking place to the ammonite form. Several species were experimenting with partially or completely uncoiled shells. Some grew into veritable giants, but most seemed to opt for a much smaller size. Were these adaptations a further sign of stressful times? I suggest that they were.

The more important changes, as I touched on earlier, are those at or near the base of the food chain – the coccoliths and foraminifera whose tiny shells of calcium carbonate accumulated in their trillions to form the chalk sediment, and the diatoms and radiolarians, with their equally minuscule shells of pure silica glass that have since been dissolved and moulded into flint nodules. The succession at Zumaya is once again informative: foraminifera, too, were on the decline before the KT boundary, although the disappearance of many species was quite abrupt at the end.

More convincingly, thanks to the Deep Sea Drilling Program and its successors, we now have more than 100 scientific boreholes that have drilled through the KT boundary beneath the ocean floor. When I was at sea on DSDP Leg 75 (Chapter 5), we drilled through an excellent and continuous succession right across the KT boundary. Although there was absolutely nothing unusual in the sedimentary record, my micropalaeontologist colleagues (those who specialize in such tiny fossil organisms) on board *Glomar Challenger* pored long and hard over the samples of microflora and microfauna, which they had painstakingly separated out from their closely spaced samples throughout this whole interval. Tantalizingly, for those of us wanting an immediate answer, their work continued in their home laboratories for some months after the cruise was over – to be absolutely sure of what they found. In short, the picture that emerged showed, in part, a rapid and stepped decline of certain species up to the KT boundary and, in part, a transitional replacement of Cretaceous with Tertiary microflora. There was no clear single extinction event.

A very similar story has been demonstrated across the globe and, with a broad spread of drilling sites, some further significant elements

FIG. 21 Photographic detail of an ammonite fossil. Sliced in half and polished, this reveals the coiled form with individual chambers separated by wavy suture lines (dark), and the complex nature of calcite cements that have caused fossilisation. This very large and dominant group of marine fossils did not survive the KT extinction. Width of view 20 cm (Photo by Claire Ashford).

have become apparent. Where the changeover in species is most abrupt, this occurs over an interval of about 100,000 years. Warm-water foraminifera declined and cool ones migrated in to take their place. The coccoliths were all warm-water species – they mainly became extinct over this same time interval. The very base of the oceanic food chain had been decimated; the results would inevitably be catastrophic. Thermal stress was again implicated as an important player.

BOLIDES AND BALLISTICS

Just as we had begun to piece together both the complexities and uncertainties of what may have happened at the end of the Mesozoic Era (Fig. 20), a new catastrophist theory burst onto the scene with almost equally devastating effect for the slow accumulation of evidence and due scientific process that had gone before. It was not entirely new as various extraterrestrial theories had been put forward previously – a supernova explosion, a sudden burst of ultraviolet radiation from the sun, a shower of comets or meteorites. But the new form of catastrophist theory that emerged from work in the late 1970s, and that was published by Luis and Walter Alvarez and their colleagues from the University of California in 1980, held that a giant asteroid (technically called a bolide), measuring 10 kilometres in diameter and weighing in excess of 4 million tonnes, struck the Earth at a velocity of around 100,000 kilometres per hour, exactly 65 million years ago. There would have been a giant impact and local devastation, great fires and floods perhaps, but the Alvarez team proposed that the main damage was done by a dust cloud thrown high up into the atmosphere such that it obscured the sun for many months. Photosynthesis came to a halt and all plant life, on land and at sea, became extinct. The dinosaurs and other creatures could no longer survive.

Their somewhat tenuous premise, with its claims of such extravagant effects, was based on the discovery that a clay-rich sediment layer at the KT boundary near Gubbio in central Italy was greatly enriched in iridium, an element known to be common in some meteorites and in outer space. This has become known as an *iridium spike*. Gubbio was about 8000 kilometres away from where the last of the dinosaurs roamed the plains of the American mid-west, and was deep beneath the central Tethyan waters accumulating a slow rain of very fine sediment and planktonic microfossils. At the time of the Alvarez publication, there was no sign of an impact crater and no other supporting evidence.

Serendipitous discovery or scandalous flight of scientific fancy? Let us investigate further. The dinosaurs were such a spectacular group of animals, much feted in popular literature from Sir Arthur Conan Doyle's *Lost World*, which certainly had me mesmerized as a child, to the hundreds of beautifully illustrated information books for child and adult alike that adorn bookshops today. The blockbuster film *Jurassic Park* further immortalized their public appeal. It is small wonder, then, that people desire equal drama at the end of their reign – surely something catastrophic must have brought about the demise of so dominant a beast. Scientists are probably little different, despite their likely protestations to the contrary. When the Alvarez bolide-impact theory hit the press, the scientific community divided quickly into two groups – the pro-bolide camp and the anti-bolide camp. A great deal of time, money and effort has since been expended on both sides but much has been learnt in the process. Understandably, but frustratingly, few people including scientists seem ready to change their cherished beliefs, at least until the evidence becomes overwhelming.

So what have we learnt? First, the hunt was on for a possible impact crater. Many new craters were found (plus even more that are not craters) and we can safely conclude that large bolide impacts are more common in Earth history than was previously known. Whereas some have occurred coincident with extinctions, certainly not all can be implicated in mass extinction events. More recently, the ghost of a giant ring-like structure, about 180 kilometres in diameter, has been seen on satellite images of parts of the Yucatan Peninsula and adjacent Gulf of Mexico. This is now known as the Chicxulub crater and is believed to date from about the right time for the KT event. It has been much heralded as *the* impact feature that all had been searching for and, if this is indeed the case, then the bolide fell to Earth somewhere in the western part of Tethys. Secondly, concentrations of iridium have been found across the KT boundary at different places around the world, together with a highly stressed type of quartz grain, known as shocked quartz, which is considered a typical feature of the stress induced at impact sites.

From the anti-bolide camp, considerable work was done on the nature and occurrence of iridium. It is now known to be a common product of volcanic eruptions, such as those of Hawaii and the basalt flows associated with hotspots and rifting. Concentration of the element – an iridium spike – does not therefore require an extra-terrestrial source. One especially large outpouring of lava, which has been carefully dated as just preceding the KT boundary event, is evidenced by 500,000 square kilometres of the Deccan Traps in India. This vast expanse of basalt is very similar in origin and scale (albeit somewhat smaller) to the Siberian Traps that are believed to have played a significant part in the great Permian extinction (Chapter 3).

Having become embroiled in the KT extinction debate, at least in a minor way, I felt that I had to visit the Deccan region myself. Walking across these stacked basalt traps of western India for the first time was enormously impressive. Although much is now largely hidden by luxuriant vegetation, oxen-ploughed farmland and the ever-bustling village life that is quintessentially rural India, where the high plateau is carved open by deep river valleys, the rusty brown columns of cooled lava-flows, one closely following upon another, appear never ending. But even with a plane ride across the Deccan region, it is impossible to fully grasp the enormous scale and prolonged duration of eruption events that must have taken place all those years ago. Somewhere in the lonely heart of the Tethys Ocean, volcanic fires raged, skies glowed red and filled with a million tonnes of ash, while the Earth shook with a violence rarely ever witnessed. Certainly this would have provided a viable alternative supply of iridium across the globe, and perhaps even the marked iridium spike at the KT boundary as found in *some* sections.

And what about the shocked quartz? Once again the anti-bolide camp has been industrious. After diligent searching, shocked quartz grains have been found in the volcanic pipes of southern Africa that provide most of the world's diamonds. Here they presumably result from extremely violent and deep-seated volcanic eruptions. Provided, therefore, that the feeder vents for the Deccan Traps extended sufficiently deep into the mantle, then shocked quartz grains may also have been blown sky-high together with dust clouds and excess iridium. Many from the anti-bolide camp quickly converted to a supervolcano catastrophist explanation for the end of the dinosaurs.

Iridium has also been found to occur naturally in seawater from which microbial limestones scavenge and concentrate it, and have thus given rise to other iridium spikes, such as one that coincides approximately with the Ordovician mass extinction 443 million years ago. In fact, almost all known elements occur in trace amounts in seawater and, of these, a number of trace metals are well known to become concentrated in particular sediments – in black shales, for example, where organic matter scavenges copper, nickel, vanadium, molybdenum and uranium among others. The concentration of trace elements, including iridium, is also achieved by a chemical process of partial dissolution of limestone after it has been buried deeply and subsurface pressure and temperature is sufficiently high. Such concentrations occur very commonly as dissolution clay seams within limestone rocks.

REALITY CHECK

Not long after the bolide-impact theory was first published and the public had been thoroughly hoodwinked by such a catastrophist explanation, I found myself working with my old friend Professor

Forese Wezel at Urbino University in Italy. Not only were we studying the mid-Cretaceous black shales of the central Apennines, which I discussed in Chapter 5, but also the overlying rock succession that spanned the KT boundary. This is superbly exposed in a number of sections, including the one along the roadside leading into Gubbio where Walter Alvarez had first encountered it, while on vacation with his family I believe. For a geologist to stand before such a portentous moment in Earth history, truly the end of an era, is almost to touch deep time itself. I could see it in the faces and disposition of my colleagues as we stood there, and I could fully understand what had so mesmerized Walter Alvarez.

However, I was there primarily to understand the nature of sediment deposition on the deep sea-floor of the Tethys, and so set about my work. What we found was a quiet mid-ocean floor where a gentle rain of tiny shell debris (mainly coccoliths and foraminifera) fell from the dramas played out amid the surface plankton. This was mixed as it settled through the water column with small amounts of silt and clay – swept in from rivers, blown in from deserts, and falling from volcanic ash clouds that periodically darkened the sky. The rate at which those sediments accumulated was torturous in the extreme, a matter of centimetres per millennium. In places, I could detect more rapid influx from a dilute turbidity current (a kind of mud-charged submarine river), or the washing by of other deep-ocean currents – neither of which had been noted by previous studies of the area. There had been a certain cyclicity of sediment input, sometimes dominated by planktonic shells and other times by clay and ash, but after deep burial and the processes of dissolution I described above, this cyclicity had become accentuated into hard bands of pinkish limestone separated by thin seams of darker red-brown clay-dissolution seams. This is typical of the Scaglia Rossa formation we were studying.

I was not, therefore, surprised by Forese's assertion that his preliminary geochemical studies had revealed the concentration of many different trace metals, including iridium, in several of the other dissolution seams he had analysed – not just the one at the KT boundary.

At this point I hastened to the nearest university library in Urbino. All good academic libraries have a series of large weighty tomes neatly ranged on their shelves, where the work done on all previous expeditions of the Deep Sea Drilling Program and its later derivatives is recorded for posterity. Sometimes they are hidden in the basement or otherwise out of the way because of their large size and the space they occupy. The first 96 DSDP Legs are reported in rather lurid turquoise-coloured volumes, a not-so-subtle reminder that they were born of Poseidon's domain, and each weighs about twice as much as a modern laptop computer. Later, the volumes come in more muted blue and lilac shades and, most recently, are nothing more than a book cover around a CD-ROM! Such is progress. Rereading the work we did on board DSDP Leg 75 across the KT boundary in the South Atlantic, then little more than a narrow arm of Tethys, I was struck by several key problems and observations that are pertinent here.

First, there is the problem of bioturbation – that is the burrowing, eating and excreting of sediment on the sea-floor by a host of animals, looking for a meal, a place to hide or a place to rest. The result is to disturb the originally neat layering and smear out tell-tale microfossils from one part to another – in this case across the KT boundary. Reworking by currents that scour the sea-floor can have a similar effect.

Second, there is the problem of dissolution or partial dissolution of fossils by chemically aggressive waters. I mentioned that this occurs after burial, as in the case of the *Scaglia Rossa* formation in Italy, but it also occurs in the deeper parts of ocean basins. As a result of general ocean circulation (the global conveyor belt model that

I referred to in Chapter 6), the deep sea contains a higher proportion of carbon dioxide than does surface water and is, in effect, weakly acidic. As the minute and delicate shells of calcareous plankton – so carefully crafted at the surface – sink deeper and deeper through the water column, they pass a chemical threshold known as the *'carbonate compensation depth'* (CCD). Below this level the shells are rapidly dissolved by acid-charged bottom waters and scarcely a trace ever reaches the sea-floor.

To these I would add a third problem, not encountered on Leg 75, but one that is typical of most sections across the KT boundary now exposed on land. That is the problem of a missing section – a gap (hiatus) in the record caused by the lack of any deposition, intermittent deposition or subsequent erosion. In this way, what may in reality have been a transitional change appears in the rock record as an abrupt event. This is the norm not the exception, and it is therefore always important that we remember the problem when interpreting the geological record.

The principal observation that I would recall here from microfossil evidence across the KT boundary drilled on Leg 75 is that '... there is much alternation of some poorly, some better preserved samples, and a transitional replacement of (Cretaceous) taxa by the newly evolving Tertiary ones.' This was from my micropalaeontology colleagues. My own description of the sediments confirms that they are a normal deep ocean suite with slow but continuous sedimentation, much bioturbation by burrowing animals, and some dissolution of microfossil shells as we approach the CCD.

SO WHAT DID HAPPEN?

I may be swimming against the tide of scientific opinion in what I am about to propose. But then I have been both surprised and disappointed to see the almost tribal approach to maintaining either one belief system or another that has occurred in the geological community, and in which the whole issue of dinosaur extinction has become mired. I have even been accused, by a reviewer of a previous book, of being '...about the only geologist left who does not believe in the asteroid impact theory'. Now this I know to be untrue!

In all honesty, we cannot but remain cautious in our interpretation of what really happened at the end of the Mesozoic Era. This has to be true because of our current inability to date events that far in the past more accurately than to within between 10,000 years (absolute best) and 700,000 years. It is also true because of the various common problems noted above that inevitably mask the precision in real datasets to which we aspire. We can, nonetheless, draw reasonable conclusions from the large amount of information now gathered by many scientists across the world.

For the bolide-impact theory, despite huge efforts and great protestations, the evidence, I would argue, remains meagre. The assumed impact crater at Chicxulub may indeed be the result of a very large, high velocity bolide, but precise dating remains elusive. All the other criteria for extraterrestrial influence – iridium spike, shocked quartz (plus osmium isotope ratios and global soot, which I have not gone into) – can be equally well explained by earth-bound processes. For the Deccan Traps supervolcano, we cannot doubt either its existence or scale, and dating does indicate that it just preceded the KT boundary event. The very latest research results I gleaned from Vincent Courtillot, Professor of Geophysics at the University of Paris, during a meeting in Bremen, Germany, just last month (2009). These indicate that the Deccan superplume event was several times larger than previously thought, and occurred in several distinct events over a 3 million year period between 67.5 and 64.5 million years ago.

However, the real problem for either of these catastrophist theories (or any others for that matter) is in explaining the actual extinction facts. For some organisms, it was a slow and painful decline; for others it was rapid – but *not* abrupt (Fig. 20). But many plants and animals were barely affected at all – most land plants, for example, survived. Most molluscs, sharks, bony fishes, placental mammals and all amphibians were also completely unscathed. And so on.

There is abundant evidence, I believe, for a combination of environmental drivers. As with the end Permian extinction (Chapter 3), there were large outpourings of lava just preceding the KT boundary event. The scale and duration of this event would have led to global environmental effects and enhanced stress for certain groups of organisms. Extremely high sea levels and a universally warm climate had bathed the late Cretaceous world in a prolonged state of benign tranquillity. This led to the removal of carbon dioxide from the oceans (and hence the atmosphere) in the calcium carbonate skeletons of marine plankton and ultimately in the chalk rocks that accumulated so widely on the sea-floor. This, in turn, would have significantly affected ocean chemistry. As sea levels fell, so temperatures dropped progressively and dramatically, a feature we can measure and even quantify from the oxygen isotope record in fossil shells. The evidence for thermal stress among different groups has shown up repeatedly in our examination of this time period. Coastal habitats diminished and land bridges opened up – the spread of diseases between animal groups, already subject to enhanced competition, would have increased.

It is difficult to avoid concluding that the KT extinction event was due to a complex combination of environmental factors that triggered extreme biological stress. This occurred over a contracted period of time, but was in no way instantaneous. Everywhere, the base of the food chain either altered or suffered greatly – the battle between angiosperms and gymnosperms on land; the fall of coral reefs and then rudist reefs in shallow seas; and the fundamental overhaul of primary producing plankton at sea. These individual turmoils spelt disaster for some and brought opportunity for others. This is but the natural backdrop to the rich history of life.

8

Portrait of the Tethys Seaway

In a white dress, vase for one barbed flower,
she floats moon-fingered, drifting,
a medusa in transparent skirts
or a narwhal beneath the split sea surface
impaled on her own horn.

In a petticoat of an unstudied red,
frilled, gill-pleated – the flounced mantle
of the sea-slug 'Spanish Dancer';
or the sillk within torn silk of rosebuds,
or the blood behind her eyelids.

From *Water Dancer* by Ayala Kingsley

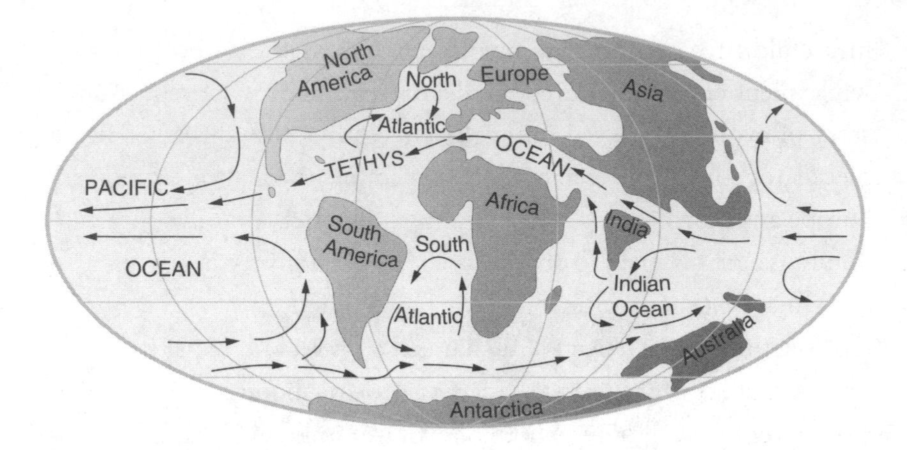

Middle Eocene Tethys Map (45 Mya). Global reconstruction with oceanic circulation, showing the much narrowed Tethys Ocean. At the same time, the North and South Atlantic Oceans and the Indian Ocean were growing in size and importance.

Roll down the years. Ring out the changes. We have now witnessed the spread of Tethys from a narrow rift across the heart of Pangaea to a mighty ocean that girdled the world. We have dwelt too long in the opulence and security of her balmy central gyres and so seen pernicious death descend. The Mesozoic has ended and now begins the Cenozoic ('modern life') Era, in which the Tethys Ocean narrowed to little more than a broad seaway between continents, as India closed in on eastern Asia, the Middle East approached Russia, and Africa ground against Europe. Already a new chain of mountains was beginning to rise – but that is for the next chapter.

Tethys was not to disappear without a final show of defiant splendour. A very different ocean world would evolve, unfamiliar at first and then progressively modern in its aspect. Monstrous shell banks glistened in the shallows, piled high with a unique single-celled organism whose disc-shaped form grew as large as the saucers in a child's tea set. Narrow ravines cut across the slopes beyond, while silent currents scoured the deep ocean floor. One remarkable story of evolution played out along the Tethys strandline has been pieced together from cetacean fossils found throughout southern Europe and North Africa. The elegance and perfection of whales and dolphins at sea is plain to anyone who has been privileged to witness these magnificent creatures in their own habitat, but fossil remains tell how they evolved from land mammals, similar perhaps to brown hyenas that scavenge along the South African shores today.

Recovery from the KT extinction was not immediate as the global environment had been dealt a near-mortal blow. It is difficult to gauge precisely how long this convalescence took – 1 to 5 million years perhaps, but the duration and nature of the recovery varied for different groups of organisms. The planktonic world had certainly suffered severe losses so that many fewer species of coccolith and only the smaller among the foraminifera were there at the beginning of the Cenozoic Era. Diatoms and dinoflagellates had not been so adversely affected. Together they re-diversified and recharged the ocean with its primary food stock. Other organisms that survived in sufficiently large numbers – bivalves, sea snails, crabs, sea urchins, bryozoans and teleost (modern) fishes – came to assume prominent ecological positions. They were soon joined by new forms adapted to particular niches, such as the flattened biscuit-sized sand dollars, which are the only sea urchins capable of burrowing into sandy beaches. Modern corals reclaimed their prime reef-building role from the rudists, although their diversification seems to have taken rather longer.

There is little doubt, therefore, that the ocean waters of Tethys provided all the vital ingredients for this renewal of life. There were abundant nutrients, dissolved gases and mineral salts. They were rich in oxygen and their surface bathed in an endless supply of sunlight. Water temperatures, though they varied slightly from place to place, were far more constant than those on the surrounding continents. But, despite these generally favourable conditions, Tethys life existed in a world of constant change and evolution, physical challenges and extreme competition. Every organism, from the single-celled master sculptors of the plankton world to giants at the pinnacle of the food chain, learnt its own way of dealing with life underwater, how to cope with patterns of light or extremes of temperature, with dissolved gases, salts and suspended sediment, with the vagaries of currents, tides and waves, and with the lottery of nutrient supply.

The adaptations contrived to meet the challenges of the marine environment were stunning in their variety and ingenuity. But every species has its optimal range of conditions for survival. Pushed beyond these limits, the organism becomes stressed, normal metabolic functions cease, and reproduction becomes increasingly difficult or impossible. Eventually the tolerance level is exceeded and the organism can no longer survive. This happened time and again in Tethys during the Cenozoic as it had before. Each extinction and new radiation brought the ocean ever closer to that which we know today.

DESERT DOLLARS

A couple of years ago I was invited to Cairo in Egypt to give a short course for the oil industry on sedimentation in the deep sea – an invitation that I was more than happy to accept. North Africa today is a unique and intriguing part of the world. Whereas Egypt may have a thriving tourist industry built up around the legendary pyramids, there as elsewhere across the breadth of the continent, lie stunning glimpses of an untouched ancient history half encased in desert sands. Phoenicians, Greeks, Etruscans and Romans have all visited, traded or lived along the northern shores. But still further back in time, and for far longer, North Africa lay on the southern margin of Tethys, either as land or sea depending on the vagaries of global sea level. There are so many hidden secrets from that more distant past that still await discovery.

So I went to Egypt with Shelley's immortal words 'My name is Ozymandias, king of kings: Look on my works, ye Mighty, and despair!' ringing in my ears, but still hoping to find signs of the Tethys that he never knew had existed. Through the windows of the modern hotel lecture suite, whenever I paused or looked out, I could see two majestic pyramids, frozen in their own time capsule, rising out of the chaotic sprawl that is Cairo today. When I visited the pyramids and Sphinx at Giza, I knew to look closely at the massive building blocks from which these iconic structures had been made. Sure enough, the blocks have been chipped and cut from the very distinctive nummulitic limestone that forms the Giza plateau and that was deposited in the Tethys Ocean between 40 and 50 million years ago.

Nummulites are small disc-shaped fossils from a few millimetres to a few centimetres in diameter. When well preserved they have a fine spiral pattern that ornamented the outer shell, but many have since been smoothed by wind and weather. Herodotus, the Greek historian and philosopher, considered their origin after visiting Egypt in the 5th century BCE, and described them as a form of lentils fed to the slaves who constructed the pyramids. A number of sacks had accidentally spilled their load, which then quickly turned to stone beneath the hot sun. Today, the Bedouins call them 'desert dollars' for they lie everywhere across the desert surface in their 'worthless' millions, worked loose from the limestone and further scattered by the elements. For palaeontologists, the truth is even more astonishing, for they were actually single-celled foraminifers – among the largest ever known – that constructed their multichambered, disc-like shells of calcium carbonate. They were a benthic species, completely different from the smaller planktonic varieties that fed on a diet of coccoliths above the Cretaceous chalk seas in a former era. They carpeted huge tracts in the warm shallows of the sea-floor and were swept in countless millions to form thick shell banks and dunes. It is from these that the nummulitic limestone lithified.

The same nummulitic limestones are found right across North Africa and the Middle East. One former PhD student of mine, Hamed Elwerfalli, traced their occurrence on the Cyrenaica Platform of northern Libya in order to map out their likely distribution below the surface, where they are known to form important reservoirs for oil and gas. Their very porous nature also makes them excellent aquifers in a region desperately short of water resources. Their distinctive signature, sometimes as glistening white cliffs more than 20 metres high, can be traced through southern Europe and across central Asia, marking out the northern shoals of Tethys.

On a different occasion, I was especially struck by a dual line of nummulitic crags just inland from the Crimean Black Sea coast that dazzled and sparkled intensely in the bright sunlight. The sparkles are actually light reflecting off tiny crystals of calcite spar, which forms a strong cement binding the individual nummulites together. My colleague and field companion at the time was Professor Roland Sobolev from the State University of Moscow. He cheerily informed me that the crags before us were the very jaws of that infamous 'valley of death' in which some British officer's blundered order had doomed the 'Charge of the Light Brigade' a century and a half before.

Some history these nummulites have seen! But they had clearly been a hugely successful organism in their own time, at their peak for about 15 million years through an especially warm phase known as the mid-Eocene high. This was a relatively short period when much elevated sea levels allowed Tethys waters once again to finger their way into northern Europe and cut across the trans-Saharan seaway as far Nigeria and hence to the widening South Atlantic Ocean, just as they had done during the late Cretaceous floods near the end of the Mesozoic Era (Chapter 7). We do not yet have a good explanation for the sudden rise in global temperature during the Eocene – but I do know several people who are working hard on the problem even as I write.

'VALLEY OF WHALES'

Having completed the short course in Cairo, I was finally able to relax with my host, another former PhD student of mine, Dr Melissa Johansson. We escaped from the constant drone of city life, the dust, bustle and smells, by hiring a felucca (the traditional sailing vessel of Egypt) and cast out on the broad and muddy Nile. The evening sky darkened over the cityscape and pyramids, our moods mellowed, and frayed edges were softened with wine, cheese and olives as we planned our excursion into the desert.

But I was in no way prepared for what I would see and learn over the next few days. We visited a Phoenician city taken by the Romans and now completely lost in the desert somewhere between Memphis and Fayoum, and half covered by sand. Roman pottery was strewn along the former streets with the same abandon as plastic bags in Cairo. There was a temple built with the same nummulitic limestone as the pyramids, now stranded half way up the Eocene hillside above Fayoum. It was here that the Egyptians first diverted the Nile to flood a whole valley and cultivate its slopes. The temple then stood near the lake shoreline. In places, where a dirt road tracked across the desert, its borders were marked by blocks and boulders – quite normal one might think, except that these boulders were actually fossilized trunks and branches from 50 million years ago when Tethys' waters rather than the Nile lapped into Fayoum Valley. We later found the fossil forest higher up the low plateau of layered rock. Desert dollars lay scattered everywhere.

Impressive yes, but not riveting. But it was into the 'Valley of Whales' that we next drove, an isolated and newly approved UNESCO World Heritage Site, which lies some 250 kilometres southwest of Cairo. I was barely able to utter a word for the rest of the day. The designated area was many square kilometres in size – and just how much more spread out into the hills and beneath the sand beyond, I do not know. The 45-million-year-old surface over which we walked was almost alive with fossils, many scattered and broken but some exceptionally well preserved. It is one of only two places known in the world (the other is in Pakistan) that hold such a fine record of the earliest cetaceans and proto-cetaceans on the planet, affording a glimpse into their evolutionary history and lifestyle, as well as just who or what they associated with.

Fifteen different species of whale have been described so far from this location; some of them, as much as 10 metres in length, are simply spread out across the desert surface today, somewhat loosely and ineffectively roped off to avoid further damage (Fig. 22). Their full skeletons, complete with ribs, backbone, skull and teeth, lie encased in a lime-rich mudstone or siltstone, preserved where they died. There are fossil turtles and sirenians (dugongs and manatees) too, a variety of small fishes, sharks' teeth and various invertebrate fossils. Partly in the same layer as most of these fossils, but more prevalent just above, are fossil mangrove roots and seagrasses. This tells us something very important about the early whales – they were near-shore creatures that lived and hunted in shallow waters around the coastline. They were also the forerunners of today's toothed whales (including dolphins and killer whales), presumably feeding mainly on a diet of fish and crustaceans. The family of baleen whales would appear much later.

From these finds in Egypt, and from finds in Pakistan that are a few million years older and so include several proto-whale fossils, as well as from other scattered fossil finds across both margins of the Tethys around this interval of time, it is possible to derive their ancestry. When the dinosaurs disappeared from land and marine reptiles from the seas at the KT boundary, there were significant ecological voids to be filled. Mammals underwent a spectacular radiation, or series of radiations, to fill all manner of terrestrial niches, which we will come to later – after all, they had been waiting in the wings for many years before, through most of the Mesozoic Era. At sea, the equivalent radiation did not occur so rapidly. Bony fish (the teleosts) grew somewhat larger and fiercer and so rose up the food chain to join the much more ancestral lineage of sharks.

It took mammals longer to respond and return to the sea, but eventually they did. The first clutch of land mammals to radiate included some altogether bizarre and cumbersome forms, which quite closely resembled some of the newly extinct dinosaurs. There was the gigantic tree-top browser *Indricotherium* of the rhinoceros family, which replaced the sauropods in Europe. Another browsing herbivore from Africa was a magnificent horned beast known as *Arsinoitherium,* quite similar to the *Triceratops* dinosaurs. This too was a fossil find from the Sahara, in slightly older rocks than those in the Valley of Whales, and is actually named after the ancient Egyptian queen Arsinoe. But the one believed to be ancestral to the whales is the immense hyena-bear called *Pachyaena,* which is thought to have been a seashore scavenger much like the modern brown hyena (or strand wolf) of South Africa. Several of the early proto-whale fossils show progressive and rapid evolution from these strand hunters to wading and walking 'whales', to paddle-swimmers and eventually

FIG. 22 Photographic detail of marine cetacean fossils (vertebrae and ribs) stretched out on the surface of Eocene sediments in the Valley of the Whales (Egypt). Width of view approximately 3 metres, individual vertebrae up to 40 cm in diameter (Photo by Erik).

into streamlined, tail-swimming, fully marine mammals. These were the true whales, such as those here in Egypt and all whales since.

OF BALEEN AND ECHO-LOCATION

Let me now follow the cetaceans and our Tethys timeline out of Egypt. They are such beautiful and intelligent creatures, whose evolution started out at a similar time to our own cousins, the ancestral primates, but could hardly be more different in the direction it took. Near Calvert in the American state of Maryland, I examined some rather younger rocks that preserve another diverse assemblage of marine mammals, including toothed whales, early dolphins, baleen whales and some primitive pinnipeds (the seal and sea lion family). These are all of Neogene age (Miocene Epoch) and represent a snapshot of evolution about 20 million years on from that at the Valley of Whales in Egypt. Tethys was much narrower by that time; its progressive closure coupled with a cooler climate and cooler seas forced a great radiation – with adaptations to new colder waters in polar regions, to freshwater habitats of estuaries and rivers, and to new types of food such as a diet of plankton.

Today, the great baleen whales are perhaps the ultimate in perfection when it comes to harvesting a meal from the ocean plankton. The blue whale, for example, the largest creature on the planet, eats an average of 4 tonnes of plankton every day. Mostly it eats krill, a small shrimp-like member of the zooplankton, which has itself fed on phytoplankton, but simply by opening its enormous mouth it swallows everything in its path. In place of teeth, these whales have great plates of baleen composed of keratin (a tough protein found in mammalian nails and hair) forming a tight mesh. When the huge mouth is full of food and water, it closes and expels the water through the baleen plates, straining out the krill and other plankton, which it then swallows. Humpback whales often feed by ascending slowly from below a rich patch of plankton blowing a ring of bubbles. This acts as a bubble net trapping the krill as a concentrated broth near the surface. Whether or not the Miocene whales had learnt the art of bubble-net fishing we cannot tell, but certainly this was the time when baleen filter feeding first evolved.

It may also have been during this Miocene radiation that the more sophisticated use of sound evolved. Certainly this is the principal information retrieval system for whales and dolphins in the present day ocean. The hauntingly beautiful songs of the humpback whales, as they migrate to and from their polar feeding grounds, have been much recorded and studied, partly with a view to gleaning some higher meaning in their communication. Dolphins have perfected the art of emitting high frequency sounds, from audible clicks to ultrasonic vibrations of over 100 kHz, and then listening for the echoes. From these they are able to build up an astonishingly accurate image of their underwater domain – the distance and direction of objects large and small, their size and shape, texture and density, the species of fish as well as their every movement. The clicks are produced from the dolphin's large forehead region, probably by recycled compressed air, and the echoes are received over a broader area around the sides of the head and lower jaw, which is asymmetrical in order to better refine the sound. A gigantic auditory lobe in the brain helps decode such a complex array of signals. All this we know from present-day species. However, we can also tell from the primitive dolphin-like species *Kentriodon* recovered in Maryland, which had a perfectly symmetrical lower skull region, that its capacity for echo-location was relatively rudimentary. Lineages derived from *Kentriodon* then gave rise to more modern-looking dolphins, with highly developed echo-location, as well as to porpoises, killer whales, belugas and narwhals.

GLOBAL CHANGE AND OCEAN CIRCULATION

For evolutionary radiation, just as for extinction, we must look to physical causes in the natural world, often to a combination of drivers or to a cause and effect sequence. It is interesting to note that the scientific community, as far as I am aware, has never evoked a catastrophist explanation to kick-start some of the dramatic periods of evolution that Earth has witnessed, but only to cause their demise. In the foregoing, I have barely touched on the scale of change and radiation that was step by step repainting the living world into a portrait we might come to recognize. I will take up the theme of these trends towards our modern world again shortly, but first want to examine some of the geological, geographic and climatic influences that almost certainly played their part.

Earth's tectonic plates were on the march, as they always are, but now the world was beginning to regroup into a more familiar aspect. The continents were starting to take on the forms we know today and moving towards their current positions, while the oceans were also changing in size and shape. As I mentioned at the outset, the Tethys Ocean was being squeezed as the Indian subcontinent pushed north, closing in on central and eastern Asia, the Middle East approached Russia, and Africa threatened Europe. Tethys ocean crust was being rapidly subducted into a line of deep submarine trenches along the northern margin of Tethys. Thus it was that this once grand ocean narrowed to little more than a broad seaway between continents, with filaments extending around India and into Europe as best they could. At its deepest it was still oceanic. When sea level was high, then larger areas of the continents were flooded and Tethys expanded (as in the mid-Palaeogene), but not for long. Northwest Africa, in particular, drifted much closer to Spain, so that only a narrow gap linked the central and western parts of Tethys – this was the first appearance of the Gibraltar gateway, which is to be of such significance in the next chapter.

Most plate-tectonic reconstructions for this period tend to call my Western Tethys the Central Atlantic Ocean. Indeed, the South Atlantic had by now widened into a fine and broad ocean in its own right, and an incipient North Atlantic was also beginning to open up. One arm of the mid-ocean ridge ran north-west between Greenland and Canada, spreading to form the Labrador Sea, while another arm ran north-east between Scotland – Norway and Greenland. The latter eventually prevailed and so the North Atlantic of today slowly opened up. At its northernmost extremity, the ocean ridge cut through to link with the Arctic Ocean.

A major mantle plume and hotspot developed in association with rifting between Greenland and Europe, just a few million years after the Deccan superplume and associated volcanic outbursts had lit up the skies above an equatorial Tethys at the KT boundary. From this younger North Atlantic hotspot, huge outpourings of lava are now seen from Greenland, across a submarine ridge through Iceland to the Faeroe Islands and Scotland. The magnificent basalt steps that form Giant's Causeway in Northern Ireland as well as those of Fingal's Cave on the Isle of Staffa are all part of this prolonged volcanic episode, which has now centred itself on Iceland.

The final piece of this all-important jigsaw puzzle began to move into place when Australia separated from Antarctica towards the end of the Eocene Period some 36 million years ago. Antarctica remained centred and isolated about the South Pole while Australia drifted north.

With plate movements as excruciatingly slow as they are, it is difficult to be certain that we have reached another defining moment in Earth history, an arrangement of plates with their continents and oceans that is just exactly right for what was to happen next. But this is about as close as we can get – 36 million years ago. It was at this point that global circulation finally changed from equatorial to interpolar (Fig. 23). Whereas warm waters had hitherto been flowing unimpeded around the broad equatorial Tethys, a fact implicated in both the Black Death (Chapter 5) and the planktonic luxuriance of chalk seas (Chapter 6), this now ceased to be the dominant pattern. Cold waters built up around the isolated Antarctic continent and, for the first time in our story thus far, ice began to form both at sea and on land during the long dark winter months. These Antarctic waters were *very* cold by any standards of the Mesozoic world before; and being cold, they became denser and so sank to the ocean floor, spreading out slowly northwards through the other ocean basins.

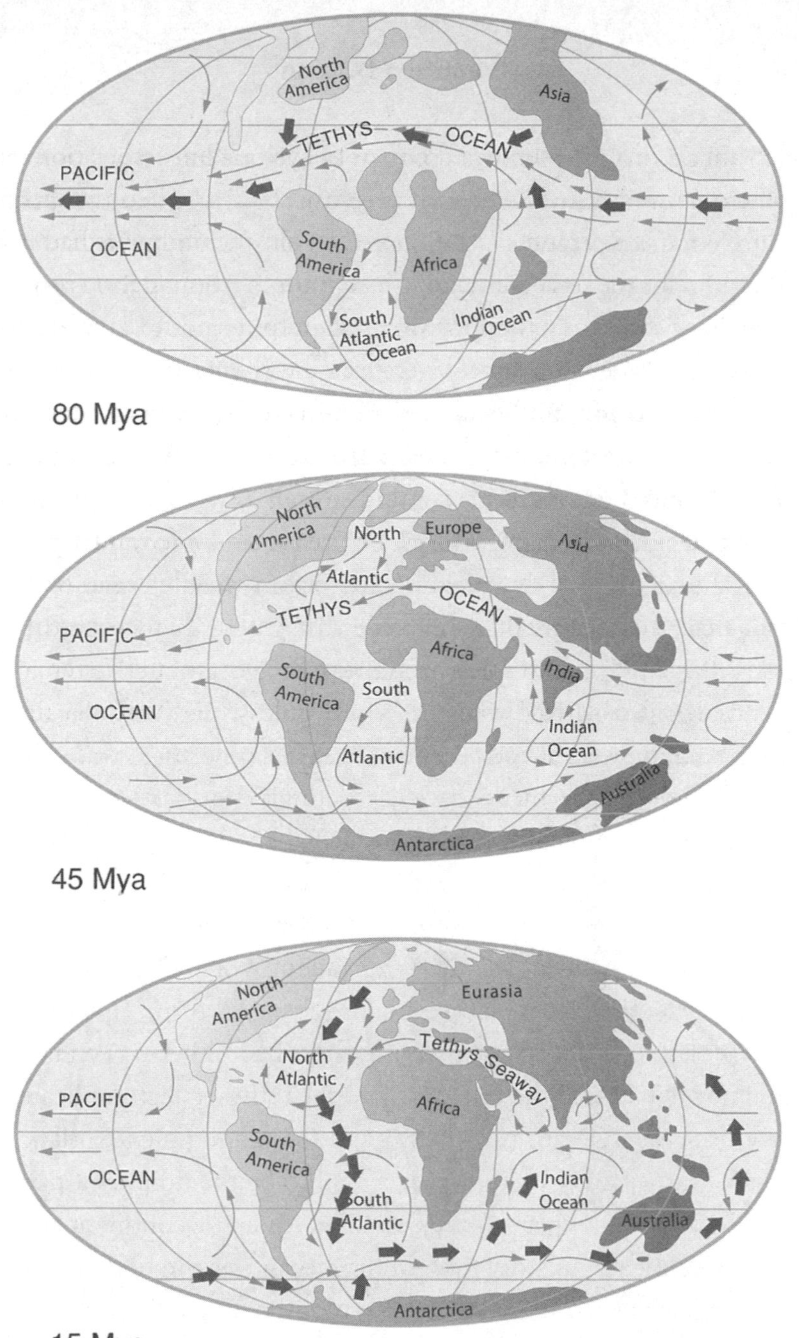

FIG. 23 Ocean circulation changes in Tethys through time. Surface circulation = long thin arrows; bottom circulation = short thick arrows. At 80 Mya there was dominant latitudinal circulation and warm bottom currents. At 15 Mya there was dominant meridional circulation and cold bottom currents. The map at 45 Mya shows a transition between these two end-member states.

It is at this point in time that oxygen isotope records, which we can construct from the fossil shells of deep-sea organisms and that act as a proxy for temperature (see Chapter 5), show a dramatic drop of nearly 5°C for the temperature of ocean bottom water in just 100,000 years. Now that really is impressively short as a geological timescale and also a very severe drop. If I were an avowed catastrophist I would be tempted to invoke the sudden input of ice-cold water into the oceans from direct impact of a monstrous comet! Not unreasonable, I feel. But no, I am perfectly content to suggest that some tipping-point had been reached in the global ocean and a 'psychrosphere' (from the Greek, *psychros*, cold) of cold bottom-water circulation was instigated. It is at this point in time that global climate began to destabilize and trend towards an impending ice age.

Although the surface waters of the Arctic Ocean and spreading North Atlantic were already connected at this time, there was still an

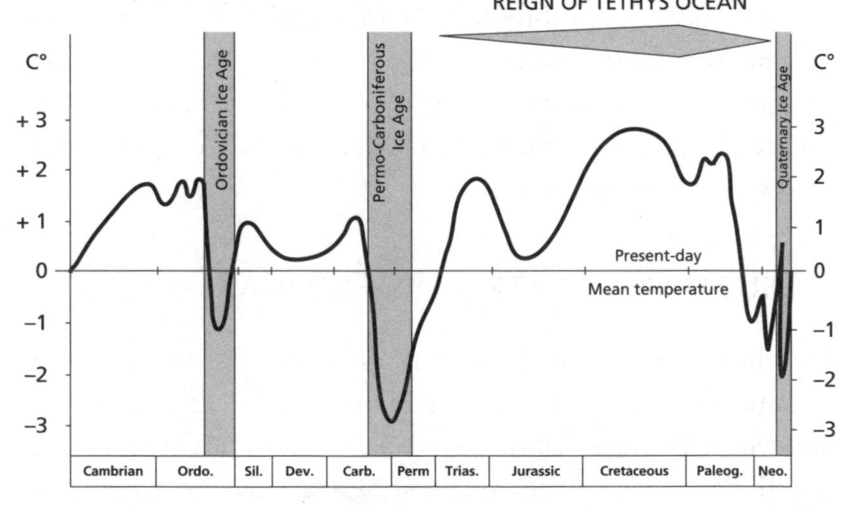

FIG. 24 Chart of temperature change through time for the past 550 million years of Earth history. Note that there have been three main periods during which time Earth experienced ice-house conditions and periodic ice ages. The normal state for Earth is for greenhouse conditions.

impediment to the exchange of cold bottom waters. The principal obstruction was the Greenland–Iceland–Scotland ridge formed by hotspot activity and outpouring of basalt lavas. Although it was entirely submarine – for even Iceland didn't appear at the surface until around 16 million years ago – it was sufficiently elevated that bottom water could not escape. We now believe that this ridge was breached about 30 million years ago, so that cold waters from the high northern latitudes further added to the psychrosphere – and the world got colder still.

Remember that we invoked these cold waters as a contributory factor in the great evolutionary radiation of cetaceans out of the Tethys. Associated with the flow of deep water are areas of upwelling in the oceans, typically along the western margins of continents in today's oceans. As mentioned earlier, these upwellings are rich in nutrient elements that feed extremely high plankton productivity, in turn attracting a feeding frenzy of fish and other animals higher up the food chain. Exactly this situation pertains at present off Peru and Namibia, for example, where it has led to vibrant fishing industries. I suggest that it was such spawning of profligate plankton blooms in the latest Eocene, available for exploitation, that led to the evolution of new species of baleen whales.

SUBMARINE GATEWAYS AND WATERFALLS

The whole world of deep-water bottom currents has been a passion of mine since the early days of my doctoral degree, when my supervisor, Professor David Piper of Dalhousie University, challenged me to find traces of their deposits on the continental slope off eastern Canada. It was then that I first learnt of the turmoil played out in the deep silence of the abyss, where massive rivers meander without banks, currents flow without end, waterfalls tumble and gurgle without sound, and storms rage unnoticed for weeks at a time. I was captivated as a PhD student, and have studied these currents ever since. We have recently found their tell-tale signature in sediments of the Tethys. So let me explain – first about the deep-water currents of today, and then those of the Tethys Ocean.

Sea-floor topography is as varied and dramatic as that on land. The large oceans are generally compartmentalized into distinct basins, with floors at quite different depths, separated by immense submarine mountain ranges. Bottom waters, generated in the cold-kitchen areas at high latitude, pile up behind such topographic barriers until they reach the spill point. Typically, this is at a narrow gateway that cuts across the barrier, just as a mountain pass weaves its way through insurmountable peaks on land. As the huge weight of dense water funnels through the narrow gateway, it is severely restricted in width and so accelerates. The bottom currents through such deep ocean passageways can be highly erosive, scouring away loose sediment, and even grinding into bare rock.

As the narrow high-velocity bottom current enters the adjoining basin, the dense water cascades down-slope and spreads out below. We call these features submarine waterfalls because of their immense height and power, although with such gentle slopes in reality they are more akin to the cataracts or rapids of terrestrial rivers. Whatever the slope, their scale is awesome. The most impressive submarine waterfall is located beneath the Denmark Strait across the Greenland–Iceland ridge. Here, 5 million cubic metres of water cascades down-slope into the North Atlantic basin *every second*, tumbling and gurgling, spawning giant eddies and turbulent whirlpools, all in the unheard silence of the deep ocean. It drops a vertical distance of over 3.5 kilometres, truly dwarfing any similar features on land. The tallest waterfall is the Angel Falls in Venezuela, which drops a mere

1 kilometre. The Cuaira Falls on the Paraguay–Brazil border have the largest average flow rate, greater than the impressive Niagara Falls – but even this is a trifling 13,000 cubic metres per second, 400 times less than its submarine counterpart.

The true power of bottom currents has only recently been realized. Once they emerge into an ocean basin, rotation of the Earth and deflection by the Coriolis force further constrain the flows against the western margin of the oceans, and transforms them into narrow, high-energy bottom currents. These are known as western boundary undercurrents, and it is these that effect the bulk of deep-water transport – water, chemicals, heat and nutrients – through the global conveyor belt. Such strong bottom currents flow for many thousands of kilometres, transporting enormous volumes of the ocean's finest sediment in addition to a dissolved cocktail of chemicals. Sediments are deposited as giant elongate mounds, known as contourite drifts, and their surfaces may be moulded into wonderfully regular waves of sediment. The contourite drifts may grow in the same place for 20 million years until they are hundreds of metres thick and cover the area of a small country – the size of Cuba, for example.

In recent years we have been gathering much more data on these unseen rivers of the silent deep and on their vast contourite deposits. The current velocity responds to small changes in climate, although the mechanisms are complex. This is because slightly more cold water is generated at high northern latitudes during somewhat intermediate climate conditions (such as today). During the extremely cold conditions of the past glacial maximum, less cold water is produced as so much is locked up as sea-ice. At much warmer temperatures, towards which climate change is fast leading us, we are also likely to see a progressive shut-down in deep-water circulation. These changes, in turn, influence the sediments deposited – slightly coarser in size at higher velocities, slightly finer at lower velocities. By coring through a thick contourite drift under the course of a known bottom current we can therefore collect a barometer of even quite subtle climate changes in the past. Working together with colleagues at the University of Southampton – Dr Sheldon Bacon and Professor Eelco Rohling – we have recently completed a large research project to do just this from the Eirik Drift off the southern tip of Greenland. Our preliminary results show bottom current velocity tracking the major climatic changes over the past 20,000 years, but with considerable variation on a shorter timescale that we have yet to decode.

But what of bottom currents and contourites in the Tethys? Remember that in the present world, deep-water circulation is inter-polar in nature and hence dominated by cold, dense waters formed at high latitudes. But for much of its existence, the Tethys maintained an equatorial circulation pattern, which contributed to a more equable global climate, when warm seas lapped the poles and palm trees fringed the northern shores of Siberia. The deep oceans were then swept by different currents, warm-water currents that formed along the margins of the ocean. Intense evaporation along arid shorelines created waters that were dense because of their excess salinity, left behind as only pure water evaporated away. These sank from the warm-water kitchens and spread out over the global ocean floor, much as the cold polar waters do today. We can be confident of this scenario as there are still two regions of the world where we can observe this phenomenon occurring today – the semi-enclosed basins of the Mediterranean and Red Seas.

Ocean modellers inform me that Tethys bottom currents would have coursed to the west, tracking with the surface flow. As the ocean narrowed during the Cenozoic Era, it became progressively compartmentalized into, at least, western, central and eastern sectors. Submerged topographic barriers between these separate basins acted as oceanic gateways through which powerful bottom currents spilled and submarine waterfalls tumbled. I have an as yet unproven hunch that just such a gateway–waterfall couplet existed near the Straits of Gibraltar between the central and western Tethys basins through much of the Cenozoic. A broad cone-shaped swathe of sediment can be seen below the present-day sea-floor in this region, sloping upwards towards Gibraltar ... but this awaits further research.

Whereas we have drilled and cored through many contourite drifts at sea, their ancient equivalents preserved in the rock record on land remain surprisingly elusive. In part, this is because there is still much scientific controversy over what contourite sediments actually look like and, in part, it is because they are rather indistinctive piles of homogeneous mud! However, having looked at very many deep-sea examples, I am confident in being able to recognize their subtle but distinctive features in 30–40 million-year-old rocks of southern Cyprus, which were then part of the Tethys ocean floor. As far as I know, these are the first well-documented examples of Tethyan contourites described – and they were discovered together with my PhD student, Dr Gisela Kahler, and Dr Costas Xenophontos from the Cyprus Geological Survey. They show a narrowing ocean and a period of intensified bottom currents – perhaps a last gasp before the equatorial circulation was finally replaced by the inter-polar circulation of the modern world.

DRAMA AMONG THE FISH

Just as the continents and oceans were realigning themselves into a more modern aspect, the patterns of ocean circulation were changing dramatically, and the world had begun its long downward spiral towards ice-house conditions, so life evolved to meet the new challenges presented. I discussed some aspects of these trends in the earlier part of this chapter – the radiation of cetaceans and the experiment with nummulites – but let me now expand further on what was happening in the Tethys, particularly with regard to the fishes. The bony fishes (modern teleost types), together with sharks, mostly sailed through the KT event completely unharmed, and so were poised to take full advantage of the new niches that had become available through extinction of other organisms.

One of the most rewarding places to visit for a portrait of Cenozoic life in the Tethys Seaway is the small village of Bolca in the Province of Verona, which can be found nestling into the verdant lower slopes of the Italian Alps. The wine-growing regions of Valpolicella and Soave lie further south, still within Verona. I would recommend a crisp dry Soave with seafood linguine in the piazza the night before, followed by a full day at the 'Pesciara' – or 'fish bowl' as the section is known locally, for obvious reasons. The village museum is spectacular.

Indeed, this 19-metre-thick limestone succession at Monte Bolca is simply packed full of the most intricately preserved fish fossils I have encountered anywhere (Fig. 25). It has been well known since the

FIG. 25 Photographic detail of beautifully preserved fish fossil from the Monte Bolca site (Italy) – width of view 25cm (Photo by Barry Marsh).

16th century, even before the time when fossils were recognized as traces of past life, so that many fanciful tales have grown up around the site. Diligent study has since revealed around 250 different species of fish from 82 families, even a crocodile, a sea snake, and a host of marine invertebrates. Within certain intervals of more oxygen-starved lime muds, the preservation is quite exceptional, so that even the internal organs are revealed and an indication of skin colouration is apparent. Everything can be found there, from tiny larval fish to large battoid stingrays over a metre in length. The diversity and abundance suggest the presence of a nearby coral reef and we certainly have much evidence from other localities that the modern hexacorals had made a full recovery after the KT event. What these sorts of finds allow is an assessment of specific evolutionary developments towards the modern world, and one of the most important for all fish was a mastery of movement in the underwater world.

Given the number of species and sheer abundance of fishes that existed at the Monte Bolca site, it is clear that they had already achieved remarkable success in evolving a range of swimming modes. It has taken millions of years of evolutionary trial and error to overcome the chief deterrent to motion through a dense medium such as water – that of drag resistance. Swimming efficiency has been achieved by minimizing the three types of drag that occur: that due to friction, turbulence and body form. To reduce surface friction, the body must be smooth and rounded. The scales of most fish are also covered with slime to further lubricate its passage through water, although we cannot assess this from the Bolca fossils. To reduce form drag, the cross-sectional area of the body should be minimal – a pencil shape would be ideal. To reduce the turbulent drag created as water flows around the moving body, a rounded front end and tapered rear is required. The resultant shape, taking into account all three types of drag, is the hydrodynamically efficient torpedo form of a tuna fish – this is the fastest swimming of all fishes today.

Sustained speed is only one of three important attributes, and is typical today of open ocean species such as tuna fish, swordfish and mackerel. A second option is that of rapid acceleration, practised by a number of reef fish that lurk in wait for their prey, perhaps beneath a rocky overhang, until their quarry is within striking distance. The third specialization is manoeuvrability, well demonstrated by the butterfly fish, for example, and the blue gill. This is used as a means of escape as well as for suddenly surprising smaller prey, and is achieved by means of the nature and arrangement of fins.

Fins play a vital and versatile role in the life of a fish, and the Monte Bolca fossils show that full use was already being made of the many different types, including dorsals along the back, caudal or tail fins, and anal fins on the belly just behind the anus. The intricate motion of these vertically oriented fins beats in consort with laterally paired sets – the pectoral and ventral fins – to yield stability and steering, forward and reverse propulsion, and braking. While the shape of the caudal fin signifies speed – lunate for fast cruising, broad and flat for acceleration – the number and delicacy of the other fins are most important for manoeuvrability.

The Monte Bolca assemblage includes a number of sharks, some of them quite similar to the tiger sharks that commonly patrol reef communities today. Sharks come from a very ancient lineage and have survived the two major mass extinction events of Tethys times, as well as others that were less dramatic. As such they have become highly evolved predators with an uncanny sense of smell, which allows them to detect food or blood in the water at levels of only a few parts per hundred million, and excellent vision up to about 15 metres. But even more remarkable, sharks together with the other elasmobranchs – skates, rays and chimeras – are also able to sense electricity, through special sensor organs located over the top and sides of the head region. These are known as the *ampullae of Lorenzini*, after the Italian marine biologist Stephano Lorenzini who first discovered them in 1678. They are small sacks filled with a jelly-like substance coating folds in highly sensitive tissue, and are connected to the surface via narrow ducts opening as tiny pores. Quite amazingly, these ampullae can detect as little as one millionth of a volt of electricity. Since all animals involuntarily generate minute electrical charges as they live and move, the ability to detect and interpret such charges is a powerful and frightening weapon. The state of preservation of the Monte Bolca sharks allows us to infer that this particular evolutionary advantage was already well established – one of the reasons why sharks have always remained at or close to the top of the marine food chain.

SEX IN THE TETHYS

What we can see through these remarkable windows into the past is Tethyan life existing in a world of evolution and change, of profusion and vitality, and of competition to survive the physical challenges that were part of everyday reality. We glimpse great ingenuity of movement for predation and escape in the fishes at Bolca, perfection in weaponry and armour among crocodiles, and the 'new' modern crabs, and even evidence for an acute ability to receive and filter sensory information among fossil sharks. But all these dramas are for nought if a creature cannot reproduce.

If we look for a moment at our oceans today, we see a scale of reproduction that is staggering. Take a child's bucket of seawater from the ocean surface and we find a teeming variety of life – most of which is invisible without a high-powered microscope. There will be over half a million unicellular plants and diatoms supporting thousands of zooplankton. During the spring bloom, when the seas become a deeper, richer green, each cell divides and reproduces so rapidly that it can yield more than one billion progeny in a month. This remarkable fecundity is almost matched by the egg-laying ability of fish and invertebrates. The average mackerel lays 100,000 eggs at a time, a hake up to a million, a haddock as many as 3 million, and a cod will even produce up to 9 million in a single egg-laying orgy. Some marine snails go further still, producing 20 million eggs, while a typical oyster lays an astonishing 500 million eggs in a single year. Sadly for the parent but happily for those who are in need of a tasty morsel, the mortality rate from such profusion is well over 99 per cent.

Although it is not possible to measure the scale of reproduction in this way for the Tethys, we can be quite certain that very little has changed. The evolution of such extreme promiscuity coupled with an elaborate richness of reproductive techniques would have been an absolute necessity from the beginning of life itself. Certainly, this would have been true throughout Tethyan history. How else would we have seen the fantastic radiation of ammonites and marine reptiles after the great Permian extinction, which brought the world to its knees? How else the see-saw evolution of elaborate reef communities – sponges, rudists, corals and their symbiotic partners? How else the phenomenal productivity of oceanic plankton that fed the Jurassic seas, caused the mid-Cretaceous Black Death and saw the spread of Chalk seas over four-fifths of the world? And now, even during the later stages of Tethys, the seas witnessed fantastic radiation of modern fish, marine mammals and much more besides.

Some 500 million years ago, before the Tethys existed, the ancient relatives of horseshoe crabs would migrate to the coast to mate and lay their eggs. Their descendants would have done so on some favoured beach along the Tethys shore (which I should dearly love to find one day), just as they continue to do to this day. Every female will lay 20,000 eggs each spring. Many open ocean fish and all kinds of bottom-dwelling crustaceans migrate towards shallow waters for breeding, especially favouring seagrass beds and mangroves. The tangle of stems and roots offer both nourishment and protection until the juvenile animals are ready to swim or crawl back to their more favoured adult habitat. It was just this kind of protective environment that is represented by the Egyptian Valley of Whales – mating and egg-laying are some of the principal reasons why the remains of so many different creatures have been preserved.

Did those early whales migrate across the Tethys Ocean like the great whales do today? Gray whales migrate some 6000 kilometres from the Bering Sea to Baja California to mate. Humpback whales turn up each spring off Hawaii for several carefree months of mating, calving and playing together, before they suddenly vanish from the waters, reappearing some weeks later to feed in the bountiful seas off Alaska. I very much doubt that the Egyptian whales strayed too far from the coast, as it was not so long before that their ancestors were scavengers of the strandline. But, by the time of their great radiation in the Miocene, when global circulation patterns changed and the cold polar oceans came into being, the new baleen whales of that age almost certainly did.

I like to think, however, that the fossil turtles probably did undertake the kind of migrations that we see today in some species of this ancient line. Green turtles that live most of their lives among the manatee seagrasses along the coast of Brazil swim halfway across the Atlantic to the tiny target of Ascension Island, more than 2000 kilometres away. The turtles mate at sea and then the female crawls up the beach to a point above the high tide line, expertly excavates a large shallow nest with her hind flippers, and lays about 100 golf-ball-sized eggs at a time. The whole exercise is repeated four or five times at twelve-day intervals, after which the adults return to their feeding grounds off Brazil, leaving the rest to fate. Remarkably, a small percentage of the hatchlings survive the entire journey as they feed and drift in the ocean currents all the way back to Brazil. When their time comes as fully mature adults, they unerringly find their way to breed on just the same beach of the same mid-ocean island – a feat of navigation whose mechanism is as yet unresolved.

There have been important and rare finds of fossil turtle eggs from the margins of the Tethys. In the Badlands of Alberta, from near the height of Tethys expansion, there is a superbly preserved female turtle with five eggs still inside, and nearby a nest with 26 fossil eggs. From the same time period as the Valley of Whales turtles, around 50 million years ago, there are fossil turtle eggs found in Eocene limestones from Alsace in France. Might this have represented migration across the Tethys Ocean? I am still looking for the former mid-Tethys island with that favoured beach.

9

Closing Ocean, Rising Mountain

'Mountain moving day has come,'
is what I say. But no one believes it.
Mountains were just sleeping for a while.
Earlier they had moved, burning with fire.
But you do not have to believe it.
O people! You'd better believe it!
All the sleeping women move
now that they awaken.

From *River of Stars* by Yosano Akiko
(translated by Sam Hamill)

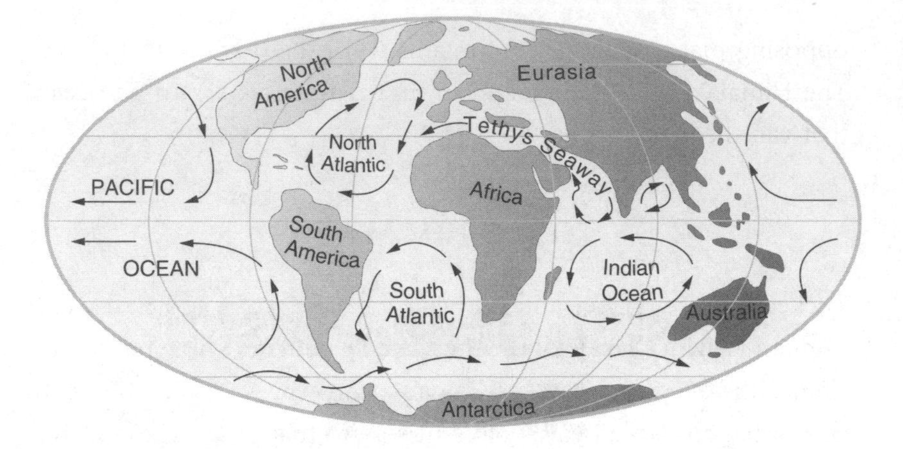

Middle Miocene Tethys map (15 Mya). Global reconstruction with oceanic circulation. The Tethys Ocean had become little more than a narrow seaway between Eurasia and Africa, and the world was fast approaching its present day geography.

Climb almost any of the mountain chains that stretch from Morocco through Europe and the Middle East to the Himalaya and you will be walking over the gnarled and folded remains of sediments once deposited on the deep floor of the Tethys Ocean or spread across the broad shelf areas at her margins. Their composition and colour, their present disposition and the scattered fossils they contain are but some of the hidden clues that allow us to recreate so vividly that splendid past. Climb still higher and you may chance upon another rock suite, with dark volcanic pillow lavas, formed at the very axis of the Tethys, the spreading ridge from which the ocean crust spewed hot and molten into those icy waters two miles below the surface. These slivers of crust from a lost ocean, some still attached to the mantle material below, are, as you will recall, known as *ophiolites*. They tell the story of how the once-broad ocean began to close, how the continents on either side ground ever closer, how the inexorable forces of collision between opposing plates squeezed mountains from the ocean into the sky. The Himalaya owe their grandeur and remote beauty to an ocean that vanished from sight.

TO THE HIMALAYA

Tucked between Tibet and the veiled nations of South East Asia, Yunnan Province is one of the most hidden and beguiling regions of China. It is the sixth largest of China's 27 provinces, similar in size to California, but with a variety and history to match anywhere I have visited. There are towering, snow-covered mountains adjoining Tibet and Burma in the north-west and lush tropical jungles that merge imperceptibly with Laos and Vietnam in the south, the border lying more or less along the line of the Tropic of Cancer. Half of all China's plant and animal species have found their own particular ecological niche in the province. It has some of the oldest hominid remains in the nation and is home still to 24 different ethnic peoples. I must have come across almost all of them while I was there, but even my field companions who spoke Mandarin Chinese often understood very little of what was being said in our efforts at conversation.

Geologically, Yunnan trends directly eastwards from the soaring Tibetan plateau and then veers southwards in a gigantic and tight tectonic arc, bearing mountainous ridges and deep valleys like the ribs of a Chinese fan. This arrangement reflects the very particular plate organization in the region, whereby India has, quite literally, rammed into Eurasia and continues to push northwards. Plunging from the 5000-metre-high Tibetan plateau are three of Asia's mightiest rivers, which have carved out immensely deep canyons in the grooves of this Chinese fan. In the renowned Three Gorges (Sanxia) area, the rivers race side by side no more than 80 kilometres apart. At one point, called Tiger Leaping Gorge (Hutiaoxia), the Yangtze River is so narrow that, as legend tells us, a hunted tiger made her escape to the other side in a single bound. There is a sheer drop of 3000 metres between mountain top and valley floor – fantastic rock exposure for a geologist, but hopelessly inaccessible! The Yangtze then trends out across the great fertile plains of central China. Of the other two rivers, the Salween exits through Burma, while the Mekong meanders more lazily through Laos, Cambodia and Vietnam. All three carry phenomenal loads of eroded mountain detritus to their flood plains and rice paddy fields and ultimately out to sea. In between the gorges lie the rocks I had come to investigate.

I was then a lecturer at Nottingham University and it had seemed like a good plan to utilize some research funds available for UK – China collaboration in order to push my investigation of the Tethys into the Himalaya. Taking up with an earlier contact I had made at a conference in Beijing, Professor Chen Changming, we arranged to meet in the far west of the province at Baoshan. I had not fully realized either the remoteness or the political sensitivity of this region close to Burma and Tibet, nor that it was normally completely off-limits to foreigners. However, due dispensation had been granted, and I was met by some of Chen's colleagues at Kunming, the beautiful lakeside capital of the province and gateway to the Stone Forest. Without them I could never have negotiated the complete anarchy at the airport as they bundled me on a plane into the unknown. In all my years of flying, before and since, I have never been quite so perturbed – by the ramshackle plane quite obviously on its very last legs (if not last flight), by the obvious overbooking as passengers crouched on their luggage in the aisle, but mostly by the 40-foot flame that seared out of the single jet engine on my side of the aircraft for a good minute or two after take-off! Against these odds, we made it to Baoshan.

The rest of the month was a veritable kaleidoscope of extraordinary images as we valiantly battled to carve a little geology out of the logistical difficulties we faced. There were market days that brimmed with produce and people in colourful costumes, but which rendered villages so chaotic we could not pass. We snaked up the steep side of a gorge through most of one baking hot day, only to find an almighty impasse at the top where a landslide had removed the road a few hours previously – while all the time lean, brown-bodied quarry workers wound up and down the narrowest of pathways cut into the vertiginous stone face with huge slabs of rock on their backs. The young lava cones by the roadside, with their bubbling hot springs and abrasively sharp volcanic road metal gave us at least three punctures – and a long waiting time while repairs were orchestrated. We came across a whole village that was a mining commune (copper and tin ore if I remember rightly) and so delighted to show us the geology in their open-cast pits – but we had to help them complete repairs to a bridge across the river before continuing. The farmer ploughing with mountain yaks, the young pig-herd in the river valley who thought my camera was a gun, the spaghetti drying stands by the roadside being turned into the sun by an old woman with bound feet . . . we encountered so many striking glimpses of life in this extraordinary region. Everyone was so helpful and generous; most had not seen a European face before. Time trickled by in rivulets of conversation, mutual interest and shared green tea.

I calculated later that we had achieved about one day in five of geology in the time we were there, geology which turned out to be both exciting and frustrating. Mountain belts anywhere, especially those that are relatively young and high like the Himalayas, only very slowly reveal their complex geology through years of painstaking research. Certain features are common to all. We found whole mountains of granite, the most common rock of continents, formed by extremes of temperature and pressure in the deep subsurface where the mix of buried rocks simply melts into a molten magma. Because hot rock is less dense, this magma then rises to a shallower depth below the mountain chain, where it crystallizes out into the very familiar minerals – glassy quartz, white or pink feldspar and shiny sparkling mica. Roadside 'factories' were busy cutting and polishing roughly quarried granite boulders into tombstones, kerbstones, kitchen worktops, floor tiles, table tops and ornaments. Grumbling trucks heaved the finished goods away, presumably destined for the burgeoning cities across the rest of China. But granite tells us nothing other than we have the exposed core of a mountain range.

Other features are rather more distinctive. We found whole sectors of mountains made of serpentinite, a dense greenish black rock with a glistening, almost watery sheen. Now this is the same rock type that I had encountered right at the beginning of my Tethyan quest in the Betic Mountains near Ronda in southern Spain. It is a highly altered mantle rock, from the deeper interior of Earth, which has been squeezed up through the overlying crust by the immense forces involved in continental collision. Here, on the Shan Plateau in western Yunnan, it is found together with disjointed slivers of lower crustal rocks (layered gabbros) and ocean-floor basalt lavas. This juxtaposition of rocks is part of the ophiolite assemblage, diagnostic of former ocean crust and mantle that has been exhumed at the surface, most typically along giant suture lines that mark the collision and grinding together of Earth's tectonic plates.

Serpentinites are a much prized rock in the building trade, as evidenced by the many stone quarries we saw in the region. But, because of their involvement in great tectonic stress during emplacement, the rocks can be much shot through with fractures. Some of these fractures are later filled by mineralizing fluids, which then paint a delicate tracery of white calcite veins through the rock while binding it tight together. Others render it too schistose (friable and flaky) to be effectively worked. We saw both types and much more besides that indicated the enormous pressures to which the whole region had been subjected.

Every geologist is familiar with the signs of faults (or fractures) in the field – polished and striated 'slickenside' faces on rocks that have slid past one another; crushed fault breccia and powdered fault gouge where the rocks have been broken and finely abraded along the line of movement; and spider-web networks of mineralized veins. Yet the number and size of faults we encountered and the overall scale of deformation across western Yunnan was beyond anything I had seen before. Rock types of entirely different sorts and ages were juxtaposed without any apparent reason. The regions in between these solid and recognizable rock outcrops were deeply eroded as valleys and gorges, sometimes displaying a truly chaotic rock type known as mélange. Indeed, the whole of the Three Gorges region is a geological suture zone that marks the welding together of two tectonic plates – that of the Indian subcontinent and the much larger Eurasian landmass.

THE FLIGHT OF INDIA

The Indian subcontinent I refer to above is a geological entity (and former tectonic plate) comprising most of what we now call India, Pakistan, Bangladesh, Sri Lanka, Nepal and Bhutan, together with parts of Afghanistan and Burma. At the beginning of this story, in the late Permian period 250 million years ago, it lay at a mid to high latitude south of the palaeo-equator, centred around 60° latitude. It was nestled somewhere between Africa–Madagascar on one side and Antarctica–Australia on the other, as part of the supercontinent Pangaea. It is worth recalling here the Permian-age rock surface with glacial striae (scratches) and other markings that I had encountered with my wife and youngest daughter in South Australia. The same evidence of Carboniferous to Permian glaciation is visible in the more southerly parts of India, Africa and South America. That was the last time, before the most recent ice age, that the Earth was caught in the grip of an icehouse climate.

Of all the dismembered continental fragments that once made up Pangaea, the Indian plate has travelled furthest and fastest (Fig. 26). It rifted from Africa and Antarctica just over 100 million years ago and started to move northwards right across the broad Tethys Ocean. In its wake, the Indian mid-ocean ridge spreading centre was hard at work, and the new Indian Ocean began to form. It was partly the development of these new ocean ridges and their high rates of spreading that have been implicated in the rise of sea level to its all-time high during late Cretaceous time, when 82% of Earth's surface was covered with water (Chapter 7). By the end of the Cretaceous it had crossed the equator and interacted with the superplume hotspot from which poured forth the basalt lavas of the Deccan Traps in west-central India. This was very probably contributory to the KT mass extinction event (Chapter 8). Around 50 to 45 Mya, it is thought that the frontal part of the Indian plate first began to collide with the Eurasian plate.

At first there were probably one or more island archipelagos stretched out along the southern margin of Eurasia. These were simply overrun, and any coastal seas they once protected were squeezed out of existence. Still the northward movement of India did not abate, although it slowed somewhat as resistance to movement increased. A range of mountains that already stretched along the coastline in this region – the ancestral Tibetan Range, I will call them – were now to receive the full onslaught as India impacted. It was this continent-to-continent collision that raised the modern

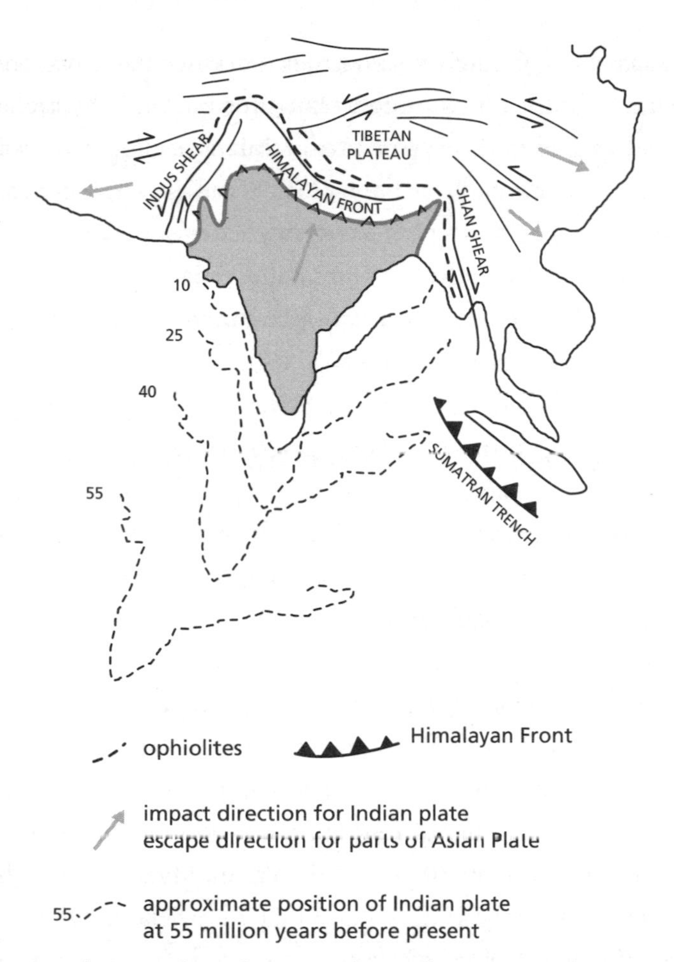

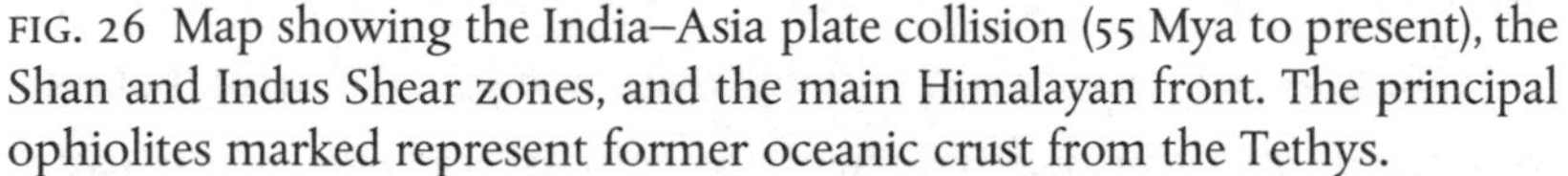

FIG. 26 Map showing the India–Asia plate collision (55 Mya to present), the Shan and Indus Shear zones, and the main Himalayan front. The principal ophiolites marked represent former oceanic crust from the Tethys.

Himalaya to their present height and grandeur, such that they now include all ten of the world's supra-8000-metre mountains. At the same time, Tibet was elevated to become the world's highest and most imposing plateau, everywhere in excess of 5000 metres high.

In total, it is possible to calculate that India moved up to 10,000 kilometres across the face of the Earth in around 100 million years.

That is an average rate of movement of 10 centimetres per year. However, it is possible to further refine our estimates by recourse to the palaeomagnetic tape recorder of ocean spreading constructed in the wake of India's flight (Chapter 6). This shows earlier rates that were considerably greater, 15–20 centimetres per year (at least as fast as the East Pacific Rise is spreading today), and rates now of 5 centimetres per year. In other words, India is still pushing north and the Himalaya are still rising.

RECYCLING AND MOUNTAIN UPLIFT

In the wake of India's flight northwards, along thousands of kilometres of mid-ocean ridge in the fledgling Indian Ocean, new ocean crust formed and spread. The same process of crustal formation has persisted for over 100 million years and still continues today, as it does in all the major oceans. This addition of material to the outer surface of the planet requires that either the Earth is steadily expanding or an equal volume of material is being destroyed annually. We now know the latter hypothesis is true. Destruction of ocean crust takes place in gargantuan recycling plants, known as subduction zones. These were introduced originally in Chapter 1, but some further explanation of the mechanisms and effects are necessary here.

Where the edges of oceanic plates collide with continents, they are subducted into oceanic trenches. The heavier oceanic crust is thrust downwards beneath lighter continental material, to be re-melted and reabsorbed into the mantle. In the upper parts of a subduction zone, melting at the surface of the descending plate occurs as the result of incredible frictional resistance. The wet sediments and rocks carried down on the plate surface, as well as those in the friction zone just above, begin to melt at depths of 50 to 100 kilometres, where temperatures have risen to between 1200 and 1500°C. Localized pockets or chambers of molten rock are created and begin to rise as great intrusions (plutons) into the lower crust where they cool and crystallize as granites – the principal rock type of the continents. Some of the molten material forces its way right to the surface, bursting out as pressure is released in highly explosive volcanic eruptions, both on land and at sea. Over the years these may build up to form whole archipelagos, known as island arcs, strung out behind the subduction zone.

The entire Pacific Ocean today is rimmed by subduction zones, trenches and island arcs and is therefore plagued with the largest number of earthquakes and volcanoes anywhere in the world. This is commonly referred to as *The Ring of Fire*. The eastern margin of the Indian Ocean plunges into a subduction zone beneath the Indonesian arc, which has experienced some of the greatest earthquakes and eruptions ever recorded. The northern margin of Tethys would have been no different. A line of trenches stretched out in an east – west direction around a quarter of the world's circumference, in places tight against the continent, elsewhere separated by an island arc and marginal sea. Everywhere along its length it would have sported a balmy subtropical climate, and the reef-rimmed islands would have been a colourful riot of life. Yet as counterpoise this would have been a veritable *Girdle of Fire* – a region of violent volcanoes and powerful earthquakes, of monstrous tsunamis that raced across the Tethys and wrought untold damage on an opposing shoreline.

But not all was quite lost. Mountains grow and so do continents – very slowly but inexorably over eons of time. The deep-seated granite melts that form from frictional heat above the subduction zone are less dense than the surrounding mantle rock and are therefore buoyant, rising up through the Earth and pushing up the land above. This emplacement of granite plutons is accompanied by the eruption of copious volumes of volcanic rock – not the dark grey-black basalt lavas of the ocean floor but lighter continental varieties (andesite, named from the Andes Mountains where it is especially abundant, as well as rhyolite, dacite and other quartz-rich types). These lavas are much more viscous than basalt and so need a much greater build-up of pressure before they can escape to the surface. The volcanic eruptions are less frequent but correspondingly far more explosive, with volcanic dust and ash, blocks and bombs spewed far and wide. And so it is that mountains grow, just like those that rimmed the northern Tethys shores.

As weathering and erosion progressively denude the rising mountains, most of the eroded sediment cascades down the steep flanks in tumultuous rivers, feeds into the ocean and is plastered against the continental margin. Little escapes beyond into true oceanic depths and even this is slowly carried back towards the continent on the spreading ocean crust. At the subduction zone, huge slices of sediment, together with parts of the ocean crust itself, are scraped onto the overriding continent, and thrust up away from the plunging slab. In the process they are badly deformed – crumpled and folded, broken apart by faults, and partly changed by extreme pressures and temperatures into metamorphic rocks. The existing rocks of the continent have themselves been placed under severe and continued pressure from the incoming oceanic plate, so that they too fold and buckle upwards. Giant rafts or blocks of oceanic and continental material, quite foreign to the local area, become fused into the growing mountain belt. They are referred to as *exotic terranes* and often remain a complete mystery as the original source may have itself been long since destroyed. And so the mountains grow still further, and thereby preserve at least some clues of a past that has otherwise vanished.

Now all the mountain growth I have been speaking about occurs on the edge of the landmass where an ocean plate plunges beneath a continental plate. Eventually, however, there comes a time when the ocean between two continental landmasses vanishes completely and continent–continent collision ensues. This is just what happened when that portion of the eastern Tethys Ocean had completely disappeared and the Indian subcontinent finally came up against the Asian continental plate (Fig. 27). Continental crust is too light and buoyant to be subducted at a trench, so the North Tethys Trench was simply overridden. The rocks of both continents buckled, folded and fractured as never before. Whole fragments of continent were detached and thrust high up over the top of the growing mountains. These superimposed rock slices are called *nappes*, from the French word for 'sheet', as they were first described by French geologists studying the Alpine Mountains of Europe.

In truth, there remains a great deal that geologists do not yet understand about the growth of the Himalaya, in large part because of their extreme complexity and almost complete inaccessibility. There are several competing theories that try to explain their great height, the elevation of the vast Tibetan plateau, their over-thickened roots, which extend far into the mantle below – almost as many as the theories for dinosaur extinction! It is possible, however, to demonstrate almost 2000 kilometres of continued penetration of India into Asia after the first tentative contact was made. As the Asian crust was pushed aside it bent and almost flowed as a great tectonic arc from the Tibetan plateau into the Three Gorges region of western Yunnan. The Shan Shear Zone is the line of fusion between plates and also the zone along which continued slippage and movement of plates still occurs.

THE BELT OF TETHYAN OPHIOLITES

As slivers of the disappearing Tethys ocean floor became plastered onto the Asian continent, or squeezed into the shear zones where the plates ground past one another, some were thrust to ever more dizzying heights. Even the 8500 m summit of Mount Makalu, a neighbour of Everest on the Nepal–Tibet border, displays a remarkable section of Tethys oceanic crust – undoubtedly the highest ophiolite in the world. Some of the sea-floor sediments deposited on this ocean crust have been pushed higher still, as evidenced by the regular layering of rock barely visible through snow-cover near the top of Mount Everest. Although a string of ultra-high-altitude ophiolites have been recorded from along the High Himalaya, as well as the sea-floor sediments associated with them, it remains logistically extremely difficult to work on them. They mark the line of ultimate closure of Tethys (Fig. 27).

One summertime, when the long winding mountain roads were just passable and the effects of the most recent landslides repaired, I found myself working in Srinagar, the lakeside capital of Jammu and Kashmir in northern India. The still lakes at dawn, their waters barely lapping the wooden sides of our houseboat hotel, were stunningly beautiful. The air was of a clarity and freshness only possible at these altitudes. Although my project at the time was concerned with landslide and related natural disasters in Kashmir, I once again managed to pick up the trail of the Tethyan ophiolite belt, some miles to the east in the Ladakh and Karakoram Himalaya: the same dark green-black serpentinites, some gabbros and pillow lava basalts, typically scattered in a chaotic mélange. They were unmistakable, and I was as intrigued by their presence as I was enchanted by the region.

Of slightly more concern, it was on this first visit to Kashmir in 1999 that I believe I came very close to being taken hostage. On the day of departure, I was met at the hotel door by two burly, bearded men who announced they would drive me to the airport, in good time for the flight because of heightened security. Indeed, there had

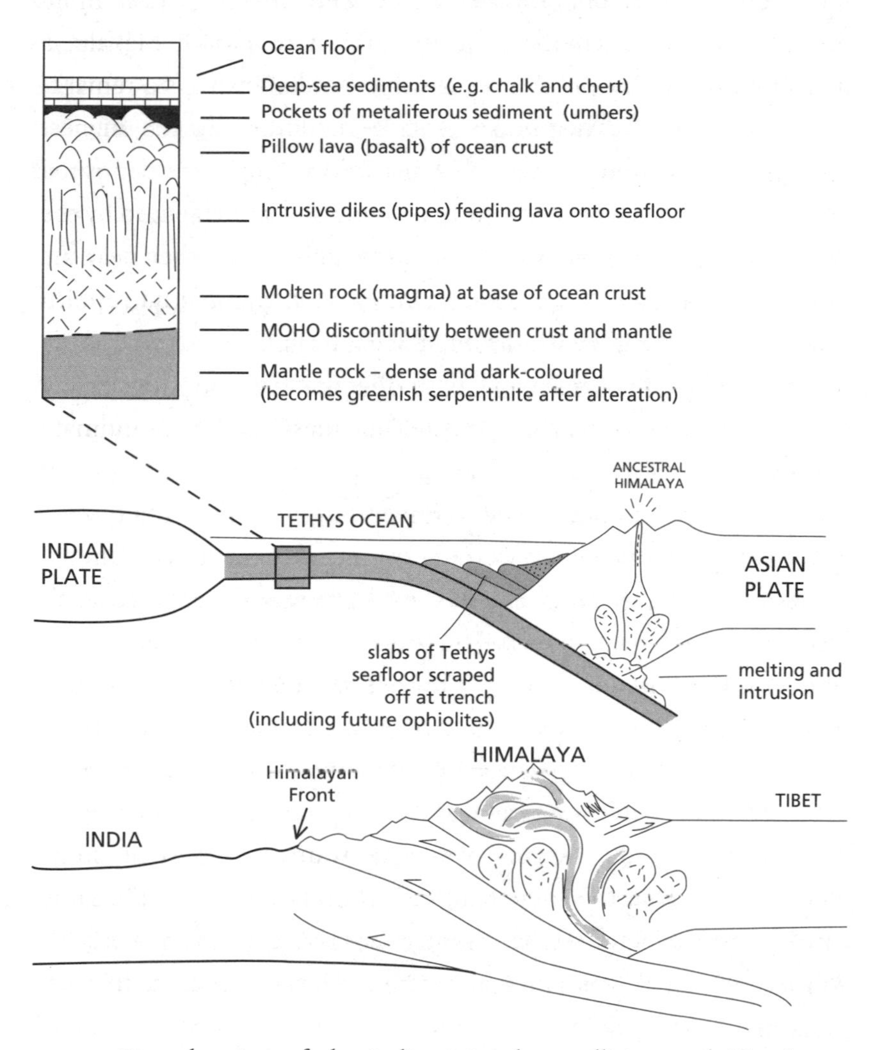

FIG. 27 Line drawing of the India–Asia plate collision and Himalayan formation. A complete ophiolite sequence is shown prior to its subduction and emplacement.

been some serious bombing incidents over the previous few days in Srinagar, and I remembered that we had been passed by literally hundreds of cavalry horses loaded onto Indian army trucks on the move to some unknown hotspot. So the caution over time seemed reasonable – although five hours before the flight seemed rather extreme, and the airport was just 15 minutes away. The men I knew as part of the group we were travelling with (though they were not geologists) and I had engaged in much interesting conversation with one of them, Rapal, about Kashmiri separatism.

After being driven for two hours directly away from the airport and in the general direction of Pakistan Kashmir, I had become a little more agitated, but still friendly in my conversation. 'We have time to show you some scenery' was all they seemed willing to tell me. Shortly afterwards, we stopped in the middle of nowhere and a third man squeezed in beside me, so that I was sandwiched in the back. I noticed the gun casually stuffed in his trouser pocket – on the opposite side to me. Soon after, we stopped again, this time in the middle of a beautifully scented pine forest – I remember that very clearly, as suddenly all my senses were working overtime. I was escorted into woods near the disputed border with Pakistan. I kept talking and asking questions, of course, and was told by the rather surly third member with the gun (the 'forest ranger' they called him) that we were tracking bears. There was absolutely nothing I could do, I had only a very general idea of where we were, and I didn't even *really* know what this was all about. Perhaps, just perhaps, it was all innocent.

The three men stopped abruptly and had a short intense conversation. It sounded angry to me, but in a Kashmiri dialect I did not know. As abruptly they turned to me: 'There will be no bears today!' they said and we returned the way we had come. I was 'released' into airport security with just enough time to catch my flight. To this day, I do not know the real purpose of this expedition nor the reason for their change of heart. I was nevertheless captivated by Kashmir's beauty and remoteness. Despite the risks, I am certainly planning a second expedition with my Asian colleagues. In some ways it is as untouched as the deep Tethys sea-floor, and ripe with untold stories.

But let me return to the ophiolites. From Kashmir, the same belt passes south through Pakistan and across the Makran – Zagros Mountain ranges of Iran. Here there is an almost straight suture zone across the heart of the Middle East, which marks where the African Plate collided with central Asia, and slivers of ophiolite and mélange escaped to the surface. The zone broadens considerably through Turkey, Syria and Cyprus into Greece and Italy, finally extending to the Ronda ophiolite of south-eastern Spain. There are literally hundreds of ophiolites to mark the passage of the Tethys, each with its own portion of ocean history to tell.

BLACK SMOKERS, TUBE WORMS AND DEEP-SEA METALS

One of the more accessible ophiolites and the first to be recognized for what it truly represented is a fragment of Tethys crust that survives on the island of Cyprus in the eastern Mediterranean Sea. Mount Olympus, the legendary home of the gods at the heart of Cyprus, is also the core of the Troodos ophiolite – about 10 square kilometres of dark, heavy and much altered mantle rock, surrounded by several hundred square kilometres of crustal rocks. It is one of the few places where you can walk along the roadside and right across the famous Moho discontinuity – the boundary between Earth's crust and mantle, which is normally buried 10–20 kilometres beneath the ocean floor.

On other parts of the island I have crossed whole hillsides of pillow lava basalts that were once ocean floor. Large depressions in the surface of some of these flows are filled with a unique, very lightweight and light-brown-coloured rock, known as *umber*. These umbers were formed from hot fluids that pumped thousands of tons of dissolved iron, manganese and other metals, which had been leached out of the ocean crust, into the cold waters on the Tethys floor. The metals precipitated out instantly as a variety of oxides and hydroxides and settled to form metal-rich sediments in natural hollows astride the ocean ridge. They have been known and extensively quarried since Roman times, for use as dyes and pigment and then later as fluxes in various industrial chemical processes. But not until much more recently did geologists begin to find the metal 'chimney stacks' that had acted as the vents for the hot fluids and their metal precipitates. Before discussing the significance of this find for the metals industry, let me first outline the nature of even more recently discovered fossil traces of the most bizarre communities of organisms that existed in these inhospitable conditions.

The finding of these fossils followed a sensational discovery in the late 1970s of two thriving communities in the present day Pacific Ocean, along the mid-ocean ridge and fracture zone close to the Galapagos Islands. Not only were these busy centres of life quite unexpected in such a barren seascape, but the creatures found there include some of the most remarkable life forms on our planet. The discovery sent shock waves through the marine biological world, with reverberations that extended far beyond. Many other such vent sites and communities have since been found, as well as their fossil equivalents. Some scientists have since proposed that such sites may hold the keys to the origin of life on Earth.

These deep-sea oases of life are centred on hydrothermal vents (i.e. submarine hot springs), which occur along mid-ocean ridges and spew hot, mineral-rich waters directly onto the ocean floor (Fig. 28). The waters mainly originate from cold seawater percolating downwards many hundreds of metres into the ocean crust where they are superheated by coming into contact with hot basaltic lava at almost 1000°C. They are further enriched in sulphur, iron, copper, zinc and other metals by exchange with the basalt, and then pumped out into near freezing seawater. The instantaneous precipitates build up tall chimney-like structures, towering above the seabed like some weird industrial wasteland. The precipitating clouds of different mineral species have resulted in the vents being called black or white smokers, and over much broader areas these clouds settle to form the umbers we see on Cyprus.

Hydrothermal vent fluids are mainly very hot (300–450°C) and rich in substances normally highly toxic to life. Yet surrounding the vents for a few metres in all directions are rich and complex marine communities. These include giant mussels, fast-growing clams, sea anemones, barnacles, limpets, amphipod crabs, worms, white shrimps and fish – most of which are unique to vents and new to science (Fig. 28). The most impressive are several new species of tubeworm, which are as thick as a human arm, up to 3 metres long and have a blood-red gill-like structure protruding from the tips of their white tubes. Pompeii worms are another variety found in cabbage-like clusters closest to the emerging water.

There is no trace of sunlight at these remote depths, so life is fuelled by chemosynthesis. Symbiotic bacteria that grow on and within the tissues of many different organisms are the primary producers, capable of oxidizing normally lethal hydrogen sulphide, and thus providing energy to manufacture organic compounds from carbon dioxide. The entire community, therefore, is based on chemical and heat energy derived from within the Earth, rather than on the external solar energy that drives photosynthesis. Some

FIG. 28 Black smoker diorama showing active chimneys constructed of metal sulphide minerals and layers of black-brown metal-rich sediment coating the deep seafloor. The strange community of organisms includes: 1 Vestimentiferan tube worms (Riftia), 2 Tubiculous polychaetes (Alvinella), 3 White shrimps, 4 White crabs, and 5 various fast-growing clams and mussels.

animals feed directly on the bacteria, such as the self-grazing shrimps, clams and mussels, while others absorb organic molecules released from the bacteria when they die. The tubeworms have symbiotic bacteria making up some 50% of their body weight and so have no need for a mouth, anus or digestive system. Clams and mussels may comprise as much as 75% bacteria, but still retain filter-feeding capability and a rudimentary digestive tract. Strange eelpout fishes nibble at the worms and clams. Not all these new species, by any means, have yet been discovered as fossils in the Tethyan ophiolites. But research is still young, and future work will undoubtedly uncover completely new species and hitherto unknown forms of life.

The other important dimension of black smokers is their metal content. For Professor Steve Scott of the University of Toronto in Canada, an ore geologist who is world-renowned for his work on these sea-floor metal deposits, the discovery of black smokers finally solved a long-standing scientific problem. This was just how and where the metal ores originated in some of the world's largest mines such as those at Rio Tinto in Spain, Noranda in Canada, and a string of giant mines along the length of the Ural Mountains in Russia. We now know that many of these deposits of mainly metal sulphides (copper, zinc, lead, silver and iron) formed as long-lived fields of black smoker chimneys and as the stockwork of feeder vents intruded into oceanic basalts and sea-floor sediments.

I recently met Steve when we were both invited to the Azores (April 2009) as keynote speakers and co-chairs of the conference session on 'mineral and hydrocarbon resources of the deep oceans'. His enthusiasm was palpable as we discussed the 350 or so black smoker metal deposits that have been discovered so far on mid-ocean ridges and in the marginal seas behind volcanic island arcs. And only 10% of the potential sites have yet been explored. This provides a potential solution to the looming crisis in supply of metals for an ever-increasing global population and its expectations for an improved standard of living. 'Take electric cars,' Steve explained, 'the vehicle of the future. Each car will require 100 kilograms of copper alone for the additional electric circuitry needed.' Seafloor mining is to become a likely reality by 2012, owing to the efforts of *Nautilus Minerals* offshore Papua New Guinea, and could well provide at least part of the answer to increased global demand.

Although we must, of course, tread with great care and responsibility in a still largely unknown environment.

Of particular interest to the Tethys story are mines in ophiolite sequences that represent the former location of active black smokers on the Tethyan sea-floor. Many millions of years later, the clouds of toxic metals they belched into those inky depths, gave rise to the mining industries of Cyprus, Turkey, Oman and other nations that straddle the Alpine–Himalayan ophiolite belt (see below). Copper, chrome, zinc, iron and silver are some of the key metals now exploited.

FROM MOUNTAIN TO DUST

The Tethyan ophiolite belt and the strange new fossil organisms that palaeontologists are beginning to find in their midst represent Tethys ocean floor. Mostly they are part of a broadly east–west system of mountain chains that result from the progressive closure of the ocean through many tens of millions of years. This has evolved from simple subduction of oceanic crust to full continent-to-continent collision. All kinds and ages of Tethyan sediments have been caught up in this prolonged mountain building episode, and it is from those that have not been too altered by heat and pressure that geologists can reconstruct so much of Tethys history. In this chapter I have mainly discussed the nature of this mountain building for the India–Asia collision and consequent disappearance of eastern Tethys.

Equally, I could have looked elsewhere along the length of the closing Tethys but, although perhaps the mountains are not as high, the story of closure and uplift in the Mediterranean region is even more complicated. As Africa pushed north into Europe and the Middle East into central Asia, it seems as though a swarm of micro-continents broke away in advance of the larger plates. It was the twists and turns of these micro-plates, fragmented at the leading edge of collision, as well as island arcs and temporary spreading centres, volcanoes and abandoned trenches, that make the jigsaw puzzle of mountain ranges and remnant basins in the whole Mediterranean region one of the most complex on Earth. My long-time colleague, Alistair Robertson, started out on his research career undertaking his doctoral degree at Leicester University on Tethyan sediments deposited on the Cyprus ophiolite. Now, as Professor of Sedimentology at Edinburgh University, he is still patiently piecing together the Mediterranean jigsaw – truly a life's work and more.

The mountain chains of Europe, North Africa and the Middle East are indeed many and varied (see map in Figure 29), including the Betics, Atlas, Pyrenees, Alps, Apennines, Dinarides, Hellenides, Carpathians, Balkans, Taurus, Caucasus and Zagros Mountains.

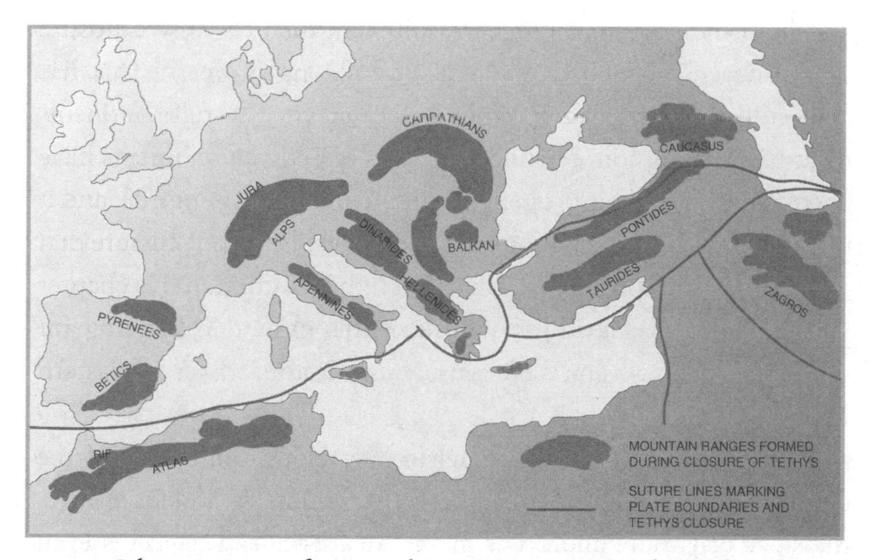

FIG. 29 Schematic map of principal mountain ranges in the Mediterranean region formed during closure of the Tethys Ocean. These are all marked by abundant ophiolites that represent the former Tethyan ocean crust.

Together with the Himalaya, they represent the most recent phase of orogeny (mountain building), which is often referred to collectively as the Alpine–Himalayan orogeny. They have all formed in some way from the turmoil of Tethys closure and plate collision in the Mediterranean region, as well as from plates grinding past one another along major suture zones. The immense traffic jam of micro-plates continues – Arabia pushes against Iran; a new ocean rift has opened up from the Gulf of Aden to the Dead Sea; Turkey is sliding west into the Aegean; while the sea-floor north of Egypt and Libya is being subducted beneath Italy, Greece and Cyprus.

The forces of plate tectonics driven by the Earth's internal engine continue to build mountain ranges and to push them high above sea level. The effects of these changes on global climate have always been marked. Before the Alpine orogeny, during late Cretaceous time, the continents covered just 18% of the world's surface area. Roll on 60–70 million years, the Tethys had all but vanished and in its place were mountain ranges and smaller seas. The total land area had almost doubled to some 32%, and the trend continued as sea level lowered into the grip of an ice age. Everywhere in the Tethys Ocean, the transition into the Cenozoic Era is seen, in terms of the sedimentary record, as a change from biogenic to clastic – that is to say, from seas dominated by the activities of organisms and the accumulation of their remains as limestone and flint, to seas receiving an ever-increasing influx of broken-up rock debris from the land.

For mountain ranges, there is a constant battle between uplift and erosion. The battle may last for tens of millions of years but the outcome is never in doubt: water will win, the broken mountain will be worn down, fragmented piece by piece into boulders, pebbles, sand and clay, and ultimately returned to the sea as dust. The mighty Himalaya have been growing for at least 50 million years and we have records on land that a proto-Indus River was already draining the western part of the young mountain range over 40 million years ago. The great fertile plains of South East Asia and eastern China have been built up over millions of years from the residue of Himalayan denudation. Beyond these, the Yellow Sea, the East China and South China Seas are all being steadily filled with sediment from the land.

The deluge of water and sediment derived from the Himalaya is staggering. I was impressed, almost frightened, by the power and fury of the trio of rivers that course through Three Gorges in Yunnan, especially while working on the rocks along their banks. But still higher in volume of flow and sediment load are the Indus and Ganges rivers, which drain the southern edge of the mountains. The Ganges alone delivers over a million tonnes of sediment each year into the Indian Ocean. It has done this, more or less, for tens of millions of years, in the process building up one of the world's largest deltas – the Ganges Delta. In fact, most of the whole nation of Bangladesh is constructed from the southward march of this delta and associated river flood plain through time – it has been some 40 million years in the making.

Of greater interest still for me, as a deep-water specialist, is the fact that the millions of tons of sediment delivered to the delta front ultimately find their way into the deep Indian Ocean beyond. The Bengal Fan, a gigantic submarine delta, is the world's single largest sediment body. It covers an area of over one million square kilometres and reaches a maximum thickness of at least 15 kilometres, thinning to a feather edge 3000 kilometres out into the Indian Ocean. In the late 1980s, Jim Cochran, of the Lamont Doherty Geological Observatory in New York, and I led an international scientific expedition to drill into the farthest part of this fan. This was part of the Ocean Drilling Program, which I have mentioned previously (Chapter 5). We joined the *Joides Resolution* drillship in Colombo, Sri Lanka, and made due preparations for the voyage. There were a number of formalities to complete with local officials, including a request from the Sri Lankan government to accommodate one of their own marine scientists on the mission. We gladly accepted Dr Wijayananda on board, who was to become a firm friend and table-tennis opponent during the months at sea. There were a final few words of encouragement for the expedition from Sir Arthur C. Clarke, whose home had been in Kandy at the centre of the island, and whose prolific and inspirational science fiction writing had been a firm favourite of mine since boyhood. And so we set sail, south beyond the equator, into the heart of that still and lonely ocean.

We drilled several sites in water depths of around 5000 metres. Our deepest hole penetrated a further 1000 metres below the sea-floor, reaching down to sediment eroded from the Himalayas some 18 million years ago. I was keen to continue deeper still and so to elucidate an even earlier record of Himalayan uplift and erosion, but time and the state of the diamond-studded drill-bit were not on our side. By analysing the composition of the entire sediment penetrated, we were able to unravel a picture of progressive denudation of the mountain range even as it was being uplifted so that today the deep core of the mountain range is exposed at its summit. We could tell this from specific mineral grains which could only have been derived from oceanic crust and mantle material.

There was much more besides that we learnt from this Indian Ocean expedition, in particular about the nature and frequency of the gigantic turbidity currents which carried their load from the Ganges Delta to the deep-sea Bengal Fan. We even discovered a completely new process of sedimentation in the deep ocean, related to the dying phases of these large submarine flows, and spent many an amicable evening sitting out under the starlit skies debating what we should call it. But I will return to turbidity currents and other deep-sea processes very shortly in the next chapter.

Interestingly, some of the mineral grains and clay particles recovered from the turbidity current deposits (turbidites) showed several additional sources of sediment besides the dominant Himalayan provenance. Some were supplied from erosion of the Deccan Traps in peninsular India which, as you will recall, were the volcanic outpourings from a superplume hotspot that burst into the middle of the Tethys Ocean 65 million years ago at the KT boundary. Some were from the warm shelf seas off Sri Lanka, and the smallest amount was derived from the nearby Afanasy Nitikin seamount islands. Many of the scientists (and even crew) searched the fine sands with extra diligence – just in case we might unearth a new concentration of Sri Lankan gems (rubies, sapphires, emeralds) in these hidden depths. But for me, the most intriguing of all was to find those distinctive dark green mineral grains of olivine, sourced from 100-million-year-old Tethys ocean crust near the top of the Himalaya and now returned to the floor of a new ocean.

10

Death Throes of an Ocean

Ah this destiny
of the darkening incessancy,
of being your own – unsculptured granite,
sheer bulk, irreducible, cold:
I was rock, dark rock
and the parting was violent,
a gash of an alien birth . . .

From *Stones of the Sky* by Pablo Neruda
(translated by James Nolan)

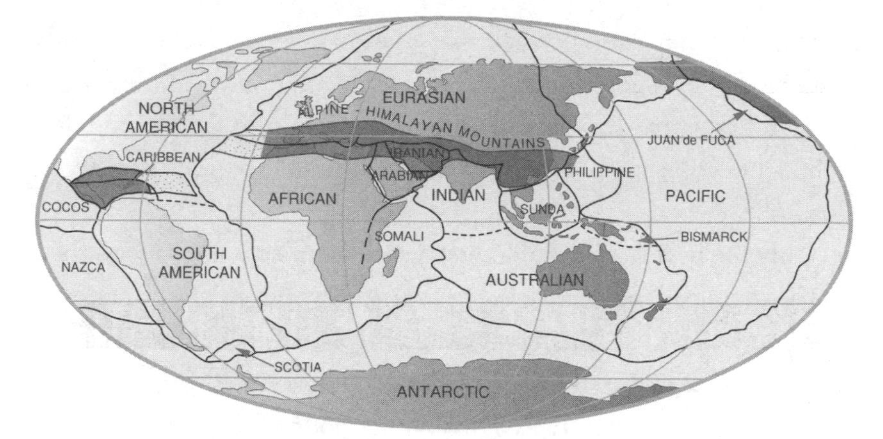

Global map showing present day distribution of principal tectonic plates. Also shown is the zone of Alpine–Himalayan mountain belts that marks the final closure of the Tethys Ocean.

Always that wide blue cloudless sky and high bright sun, but still a light spring heat and gently wafting sea breeze. Everywhere a floral cacophony of colour and scent, from pale field iris to dazzling yellow asters, white Tunisian wild rose to rich velvet trumpet flora – and many others whose names confound. Pine-wooded mountain slopes blend into rolling hills that support a thriving cork-oak industry; olive groves, carpeted with waves of poppies, are carefully managed for export to Europe; while the lowland meadows are a patchwork of barley fields, rustling and alive in the wind, vivid yellow oilseed rape and freshly ploughed rich soils.

I have come to Tunisia to write this last chapter, where I know a fine and friendly hotel, serving good wine and seafood. Imitation French colonial in style and perhaps a little ramshackle in places, but it lies just above the small coastal town of Tabarka and has a spacious terrace area with pool, and shade for writing – and a magnificent vista across the Mediterranean Sea. Already the pace of time has slowed. Donkeys laden with leafy vegetables plod sullenly and stubbornly from nowhere to somewhere; a straggle of goats and sheep sprawl across the bumpy rough track that masquerades as a road, carefully shepherded by a colourfully clad Berber woman; men in their worn Arab robes sit by the roadside in contemplation of the passing day, while the womenfolk walk for miles bent double beneath some vast load of firewood.

Only the port is a hive of industry as fishermen coil or mend their nets and offload their catch, buyers carry away bright plastic crates of glistening silver fish, diving into the narrow streets behind, weaving their tortuous way to the fish market. I love the smells from this part of the souk, as well as the noise and clamour of the adjoining part where fruit and vegetables are piled high, and baskets brim with all kinds of pulses and spices. But I walk quickly through the meat section, where whole heads and carcasses are hung out for show, all kinds of innards extracted and on view, and patches of deep red blood stain the slabs of the narrow street.

This is but procrastination. There are at least two important reasons for being here in Tunisia to finish the book. First, below my feet where I sit writing and across the bay where the old Citadel perches are deep-sea sandstones deposited from the mountains that rose up as Tethys closed. Second, the azure Mediterranean beyond has swept in to fill the gaping basin and cover the blistering white salt flats that were left behind when the last of the Tethys waters finally evaporated.

CANYONS AND CATASTROPHES

The imposing sandstone cliffs at Tabarka tell a remarkable tale of the rise and fall of mountains and changing allegiance between continents. The young Alpine chains along both shores of Tethys were beset by earthquakes and volcanoes, and then savagely eroded, feeding into great rivers that discharged their detritus back into the ocean. This material was then transferred to the deep sea by submarine slides, debris flows and giant turbidity currents that coursed down great canyons and channels cut into the sea-floor (Fig. 30). These are the sorts of sediments we drilled on the Bengal Fan (Chapter 9). The rocks along the northern coast of Tunisia form part of a 2000-kilometre-long belt, known as the Numidian Flysch, that stretches from Calabria in southern Italy, through Sicily, Tunisia, Algeria and Morocco and curves across Gibraltar into southern Spain.

I have been studying the Numidian Flysch for some years now, examining the sandy fill of deep submarine ravines that formerly scoured the Tethys slopes between 30 and 40 million years ago, and trying to understand just how the thickest and completely structureless sands had been deposited. Near the beginning of this programme in 1992, I was leading a research workshop on the problem in the delightful coastal town of Cefalu in Sicily, and stood with some 50

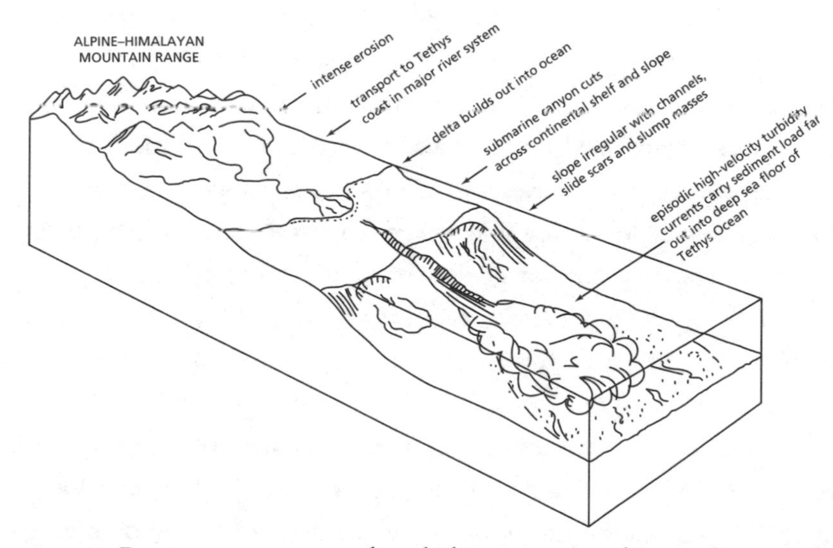

FIG. 30 Deep-sea canyons and turbidity currents schema, showing the route by which sediment eroded from rising mountains found its way into the deep Tethys Ocean.

other deep-sea specialists in one large ancient canyon now exposed along the shore. As we were reflecting on the scale of the process whose sandstone deposit now dwarfed us, Claire (my non-geologist wife) proposed an interesting idea to do with sand and water sloshing around in a child's bucket. Many years later, and with some refinements, this is exactly what we believe to be the best explanation for the process that causes these deep-water massive sands.

Some of the canyons that cut across the Tethys slope were impressive features, incised hundreds of metres into the sediment, several kilometres in width and hundreds of kilometres long – like Grand Canyons of the deep. We can gauge the nature of the turbidity currents that raced down their length from the deposits they left behind, called turbidites, and also from indirect observations of their counterparts in today's oceans. Essentially, they are very large sediment-charged flows, much like a large turbulent river in flood, that run in channels across the sea-floor far below the surface. They can completely fill the channels through which they flow and overtop the margins, just as a river breaches its levees and spills over a floodplain. On steeper slopes they can reach speeds of over 80 km/h (50 mph) carrying pebbles, sand and a large amount of mud.

Other parts of the Tethys slope were steep, irregular and subjected to considerable instability and down-slope movement. This often occurs where large amounts of unconsolidated sediment have piled up rapidly on the outer shelf edge from rivers and their deltas, or in areas that are especially prone to earthquakes. Both of these 'trigger mechanisms' for slides would have been common along the Tethys margins at this time, with the resultant displacement of huge volumes of material. Evidence from recent examples of such catastrophic slides at sea have involved the sudden disappearance of an oil-drilling platform off the Mississippi Delta, the collapse of half the new runway at Nice Airport in southern France and its removal down-slope into the deep Mediterranean, and the overnight vanishing of a Ukrainian village on the Crimean Peninsula, as it slid towards a watery grave in the Black Sea. The onshore equivalent in scale for one of the larger slides would be moving the whole of Greater London sitting on a thick slab of sediment across the Channel into northern France.

SAND GRAINS AND CRYSTALS

Ever since I first read William Blake's perceptive words: 'To see the world in a grain of sand, | heaven in a wild flower; | hold infinity in the palm of your hand, | eternity in an hour,' I have adopted them as a creed for geologists. This is more or less just what we aspire to. Later, when I met Paul Potter, Professor of Geology at the University of Cincinnati in the USA, I much admired the sheer audacity of his early sedimentological work, which mainly involved sitting on as many beaches as he was able to visit around the coastline of South America. He then wrote serious scientific papers about the nature of the continental interior based on analysis of the sand grains from every beach. His work was much admired by many throughout South America too, for we next met as invited guests of honour at the Chilean Geological Congress of 1991 held in Valparaiso. My keynote address was on the deep-sea I had never visited, except via remote surveying; while his was on the many beach sands in his collection!

Incidentally, it was here that I first came across the inspirational Chilean poet, Pablo Neruda, and his collection of works 'Stones of the Sky' (*Las Piedros del Cielo*, translated by James Nolan). His ability to capture the meaning and beauty of Earth, her rocks and minerals, is second to none, and so it is from this slim volume of Neruda's poetry that many of my chapter openers have been chosen.

But, back to Tunisia. Most recently I have been working with Professor Mohamed Soussi from the Faculty of Science at the University of Tunis, and with two of our joint doctoral students, Sami Riahi and Chrissie Fildes (who do most of the work, of course!). What we have discovered, hidden amid the rose-coloured quartz grains that dominate the old and sea-battered rocks at Tabarka, are a very few, perfectly formed, zircon crystals – each no larger than a pinhead. Separated from their host sandstone by means of a hammer and ultrasonic bath, and then using sophisticated radiometric dating techniques these zircon grains have yielded ages of just over 500 million years.

Now, this relatively 'young' age allows us to resolve a long-standing debate among geologists, of just where these rocks have come from. It confirms that a whole swathe of northern Africa, from Tunisia through Algeria to Morocco, has much closer affinities with a *younger* Europe than with a generally *older* Africa. Indeed, we can deduce that the Numidian Flysch was once firmly attached most probably to a part of Spain and France that now lies over 800 kilometres away from Tunisia on the other side of the Mediterranean.

But how can we determine that such dramatic changes have taken place in the face of our world from a single grain of sand? A not unreasonable question. In fact, there are many telltale signs now hidden in these sands and intervening muds that record their original deposition somewhere deep within that vanished ocean. The first among these is the fact that they are a whole stack of turbidites, associated with slides, slumps and submarine debris flow deposits. Characteristic microfossils recovered from the muddy sediments in between sandstone layers allowed us to date their origin to between 20 and 30 million years ago, but still we did not know from which land the grains of sand and tiny particles of mud were eroded and transported seawards.

It is routine practice for geologists to take samples from the field and cut them into very thin slices, using a diamond-studded saw back in the laboratory. These slices are then stuck onto small glass plates and ground into ultra-thin layers that are semi-transparent under the microscope. It was in one of these 'thin-sections' of rock that we first saw the tiny scattered grains of zircon.

This in itself was not unexpected as zircon, like quartz, is a highly durable mineral, resistant to many cycles of weathering, erosion and transport. What was interesting, however, is that when zircon is first formed, somewhere very deep and hot within the bowels of a mountain chain, it locks trace amounts of radioactive uranium into its crystal lattice. Uranium decays to the stable element lead at a constant and extremely slow rate so that it can be used as a radiometric clock to determine the age at which the zircon crystallized from its molten precursor. This is done by accurate micro-measurement of the relative amounts of uranium and lead in an individual grain and by knowing that the time it takes for half the uranium (235 isotope) to change to lead (207 isotope) is 713 million years.

It was a 'eureka' moment when the machine finally churned out the results from our Tunisian zircons. Every one we measured came out with an age somewhere between 500 and 520 million years. Even with that degree of accuracy, there was no escaping the fact that these dates correspond well with some early important mountain building periods in Europe, which is where our zircons were almost certainly formed. As these mountains were eroded, including the granites in which the zircons initially crystallized, they shed their mineral load to rivers and eventually the sea.

Perhaps there were even several phases of uplift and erosion before the tiny zircon grains ended up on the deep sea-floor of Tethys Ocean somewhere south of Europe. When the Tethys Ocean finally vanished between the gigantic onslaught of plate tectonic collision, parts of the ocean floor were scraped up onto land and thrust over the continent as 'nappes' – where they now form part of the Numidian Flysch that outcrops in Tabarka and all across northern Africa.

So why not an African origin in the first place? Very simple. The age of similar events on the African continent are all at least 2000 million years old and would therefore have yielded much older zircons than we found in the Tabarka sands. Partly in order to ensure that this last statement was true, we also analysed zircons from sands we collected further south in Tunisia, near Testour. These were from a well-known fluvial succession, also around 30 million years old, which occurs over large tracts of North Africa. This Nubian sandstone does indeed represent the normal river drainage from the African plate to the Tethys Ocean. Sure enough, all these zircons gave a date close to 2000 million years.

LIFE IN TURBULENT TIMES

Minerals that started life as tiny crystals in granite on one continental plate and then ended up, via a spell at the bottom of the Tethys Ocean, as sand grains on a different continent; slices of the continental slope the size of Greater London that slide, under the force of gravity, to the bottom of the ocean, detached by some especially virulent earthquake tremor; whole rock formations that are thrust up and over others as they build up into mountain ranges; fragments of the ocean crust and mantle that are squeezed up along suture zones to the very highest parts of the Himalaya – these are but the norm for the turbulent times which brought Tethys to an end. Many of them were played out in geological time, over a few tens of millions of years. Other events are almost instantaneous, such as submarine slides and turbidity currents, or the earthquakes that triggered them.

As a backdrop to the canvas of life, this was little different from the world today. At that time, Earth was plunging towards an ice age, while we currently seem hell bent on catapulting ourselves away from one. The Tethys had not yet taken its last gasp. Otherwise, almost all the physical parameters were similar. It is no surprise, therefore, that most of the fossils we find from the early Neogene period (Miocene Epoch, between 24 and 5 Mya) are almost identical to living forms today. Both the oceans and the land were rapidly adopting a thoroughly modern aspect.

Miocene sediments are abundant below many parts of the seafloor and have been extensively cored on very many scientific drilling expeditions. Because of the continuing tectonic unrest along the Tethys margins, Miocene rocks are also well exposed on land. They are often quite unconsolidated, so that fossils simply fall from them with little effort on the part of the collector, although elsewhere they are rather more hardened from the trials of burial, heat and pressure. Both conditions are true of a thick rock succession in south-eastern Spain, not far from where I first started to write Chapter 1. If I peer through the spectacular sunset from my open-air terrace-office in Tabarka, I know that I am looking towards the small unpretentious town of Carboneras, somewhere well over the horizon.

Carboneras is a working town not far from the over-popular tourist trap at Mojacar. I once saw it translated literally on an English website as 'Coal Bunkers' – in a vain effort to attract tourists away from its better-known neighbour. In fact, there *are* a growing number of Spanish tourists who come to escape the August heat of Madrid, Cordoba or Seville, and a very pleasant new seaside promenade has been constructed. The town boasts a small but thriving port, several dusty limestone quarries with gargantuan trucks constantly feeding an imposing cement factory at the edge of the town, and an exciting new desalination plant, which had just been commissioned when I last visited. On the dry plains all around there are enormous plastic greenhouses, covering hectare upon hectare of land, where the vegetables for the rest of Spain and much of Europe are grown. Their dazzling reflectivity is the first feature of scenery visible on flying into Almeria. The coastal range of hills to the south is part of the eerily beautiful Cabo de Gata national park. Carboneras is the perfect place from which to explore some classic geology near the end of the Tethys history, followed by a good Rioja in a Spanish tapas bar.

Stretching from just north of the town, across the plains beneath those ubiquitous plastic constructions, and to Almeria Bay – a distance of some 30 kilometres – there lies the Carboneras 'Rainbow' Fault Zone. This is truly the San Andreas Fault of southern Europe, which lies like a giant's multicoloured cloak or scarf, carelessly draped across the landscape, valley and mountainside alike. The different colours – red, brown, green, yellow, grey and black – are formed of different rock types that have gouged and mixed and strung out along the boundary between two continental plates as they ground past one another. When I first came across this impressive but strangely sinister zone I was on holiday with my older children, Jay and Lani, who were still only 9 and 6 years old at the time. I have a memorable photograph of the two of them posing in their colourful summer outfits on part of the Rainbow Fault. Both of them were distraught to find out later that they were simply there as a very small scale to show a very large fault zone that is as much as 2 kilometres wide in places – and also a little perturbed when I told them that it was still moving, albeit at an average rate of 0.2 millimetres per year – enough to cause devastating earthquakes in the recent past at Almeria, Sorbas and in other towns that lie on the fault or one of its many offshoots.

Carboneras and the rugged landscape of the Cabo de Gata peninsula lie on the Mediterranean side of the Rainbow Fault. The domed hills are the remains of Miocene volcanic seamounts that formerly lay somewhere off the African coastline, spurting dust and ash high into the atmosphere. Rondan Volcano, just past the cement factory chimney, is capped with a coral reef, and fossils identical to reef faunas today, while coastal and lagoonal limestones are spread about its base. The fossil oysters seem so fresh as to be almost edible – although they too prove to be around 10 million years old. Rodalquilar Volcano, a few kilometres down the coast, has been worked through the years as a very productive gold mine, although it is now being turned into a mining museum. The hot fluids rising up through these volcanic rocks, as well as those permeating the Rainbow Fault, have brought gold near to the surface; and there is still active prospecting in the region. Los Frailes Volcano sports a 5-kilometre-wide crater, half of which has been lost to the sea, which was blown out in a final catastrophic act of defiance about 14 million years ago. The fine white ash-fall is still seen patchily for miles around, in places quarried as bentonite clay, and in others fallen as Miocene beach sands that are now crumbling into the beaches of today. The seashore fossil shells mix almost imperceptibly with their modern counterparts.

On the other side of the Rainbow Fault, life along the Tethys shores of Europe was much the same: fewer volcanoes, extensive coral reefs that rimmed the continent, thriving fish and plankton out to sea. Subtle but important changes are noted from this time, such as the first appearance of robust algal ridges on coral reefs. This gave them the capacity to form strong and substantial barriers facing powerful waves. Since Miocene times, coral reefs across the world have been protected by algal ridges, and have therefore come to thrive along coasts pounded by heavy surf. However, as Charles Darwin had observed a century and a half ago, invertebrate life tends to evolve less rapidly than vertebrate life. Only modest evolutionary changes among the marine invertebrates separate the Miocene from today.

THE MESSEL PIT

While the portrait I have painted of the Cenozoic Era thus far has been about the slow but inevitable closure of Tethys, the growth of mountains and the almost doubling of land areas, I have mentioned little of life on the land that abutted the Tethys Ocean. This is important, not least because of our own more personal involvement as a species. I will first backtrack, therefore, to a time earlier in the Cenozoic, when creatures akin to the strand wolves of South Africa took to the seas again and started down an evolutionary pathway towards modern whales and dolphins – we saw the start of this at the Valley of Whales in Egypt (Chapter 8). There is another hugely significant UNESCO World Heritage Site at Messel in southern Germany, which preserves in exquisite detail a snapshot of life on land at exactly the same period, some 48 million years ago. To say that the Messel Pit is rather less prepossessing than the desert site in Egypt would be a massive understatement, but one should not be deceived by outward appearances.

Messel is a small, rather faded industrial town just 30 kilometres south of the modern and thriving city of Frankfurt. Amazingly, despite having spent many happy childhood exchanges with a family in Frankfurt, I never once heard any mention of Messel. Most locals still know nothing of *Messel Grube* (the Messel Pit), although its recent rise to notoriety in 2009 through the unveiling of *Ida* – an almost prefect fossil primate found at the site – may have done something to raise awareness of this unique locality. *Ida* is formally known as *Darwinius masillae,* and has been much heralded in the media as the missing link to our former ancestors. This is not in fact true. Although it is a uniquely complete specimen, other similar primate fossils are known and this one is most probably from an undistinguished sideline.

Tucked behind an industrial estate is a blackened gaping hole with a small pond tending to putrefaction in the heat, nonetheless alive with loud croaking frogs. This is it: the site of what was formerly a small deep volcanic lake near the edge of Tethys, with noxious fumes and sulphurous air at the surface and even more putrid bottom waters than today. But it was surrounded by a lush subtropical forest in its Eocene heyday, and acted as a magnet for life at that time. Thirty-five different species of mammal have been found in the Messel Pit so far – bats, insectivores, carnivores, ungulates, anteaters, marsupials, rodents and primates. Where to start? There is a miniature horse, pregnant with foal; a bat whose stomach contents show its favourite snack was a butterfly; a strange gerbil-like insectivore, unlike anything we know today; primitive hedgehogs; and lemur-like primates including *Ida*, who were already skilled tree climbers.

It was not only mammals that were drawn to their untimely end at Messel. Several pairs of turtles have been found, fossilized almost in the act of copulation; crocodiles pursuing perch and other freshwater fish; and even birds that somehow flew too close to the dark alluring water for comfort. My own favourites are the insects – queen ants with a 12-centimetre wingspan; large ungainly cockroaches whose form has been successful since it first appeared; and beetles whose patina of colours have been miraculously preserved – iridescent orange and metallic blue-green. These insects and their colours were preserved as they fell through the lake waters, came to rest between layers of soft black organic-rich mud, turned slowly to stone, and then survived buried for 48 million years. There have been other more recent close shaves for the Messel fossils, particularly when mining oil shales became fashionable in the 19th century, when the anarchic gold-rush of fossil hunters moved in, and when plans to turn the whole area into a landfill site were very nearly realized. Being a perfect fossil is a precarious life indeed.

COLOURFUL GRASSLANDS

The early Cenozoic Era was an almost modern world – but not quite. Although most of the modern lines had been established, there was a serious extinction event at the end of Eocene time, coincident with a dramatic fall in mean global temperature. A number of the more unusual forms disappeared at this point. Prolific evolutionary radiations once again filled the vacant niches and adapted to a new world, which was both much cooler and significantly drier as well. The spread of new desert areas and semi-arid lands was probably of equal significance to the global cooling trend, for the changes that came about in the terrestrial floras of the Tethyan region, and which spread globally. Both were also germane in what was later to befall Tethys.

Lush subtropical forests such as those surrounding the Messel Pit, where fossils of laurel, oak, beech, citrus, vines, palms and water lilies have been found, became much more restricted in their extent. The same was true of the equatorial flora. In their place, grasses first appeared and underwent a massive radiation in diversity and abundance. With the greater climate differential on Earth, the winds became generally stronger and more persistent, significantly aiding in the distribution of pollen to new areas. Accompanying the grasses were herbaceous plants (weeds) of all kinds. The *Compositae* family (including asters and daisies), whose bright colours first caught my eye when I arrived in Tunisia, evolved at this time. These and other herbaceous weeds became especially rife. Their pollination was greatly aided by a burgeoning insect population.

These widespread changes at the base of the food chain on land had huge ramifications in the animal world. At first it seems that savannahs proliferated (these have scattered bushes and trees), and then the prairies and steppes (without any trees). In both areas, large herbivores appeared and radiated, grazing on the grasses, browsing the bushes and trees. These were mainly the ungulates that are so familiar today – cows, sheep, goats, deer, giraffes, camels and llamas (all with an even number of toes), and horses, zebras, tapirs and rhinoceroses (with an odd number of toes). They developed digestive tracts to cope with the new fibrous diet, as well as strategies to avoid excessive predation. The latter seemed to involve either speed of flight or a herding instinct, or both. The principal predators that evolved at the top of this rather short food chain were the cat and dog families.

Much smaller herbivores and related animals also evolved at this time in response to the changing vegetation. These included the rabbit group and the rodents – rats, mice, moles, voles, lemmings, and some others – that developed specialist diets, including specific grasses, weeds and seeds in particular. They also needed effective strategies to cope with the larger number of predators that their small size, and hence vulnerability, attracted: their speed of movement; generally being hidden below the lower leaves of plants; the nocturnal or semi-nocturnal lifestyle of some; and the great innovation of burrowing for safety. Rabbits and rodents also evolved very high rates of reproduction; 'breeding like rabbits' is the expression, although mice are even more prolific in this way. There is also much truth in the veiled warning that we live no more than two metres from the nearest family of rats.

It is salutary to reflect that so many groups of animals, including primates and ultimately humans, have had their evolutionary pathways affected by the emergence and spread of grasslands across the world. The nature and health of species at the base of food chains, both on land and at sea, have the most profound effects on all other organisms – for evolutionary development and radiation, as well as for extinction events.

PRIMATE EVOLUTION

Did our distant human ancestors ever see the Tethys Ocean before it vanished? This is an interesting question, but a very difficult one to answer. Because primate fossils are not especially common, the evidence for their evolution and radiation is frustratingly patchy – a tooth here, a bone there, and long intervals of time in between. Primates probably first evolved from small insectivorous mammals, not unlike today's shrews, which played hide-and-seek with the dominant dinosaurs of the day. This was some time during the Cretaceous period when Tethys was at or near her maximum extent and the world was the warmest it had been for many millennia. Fortunately for us, they survived the KT extinction event and, by the early Cenozoic Era they had split into at least two groups: one was lemur-like (*Ida* from Messel fits best into this category) and the other tarsier-like, small agile animals living mainly in trees on a diet of fruit and insects. It is from this second group that first monkeys and then apes descended.

The earliest apes were forest dwellers living in tropical East Africa. They had a thumb that could grip and spent much of their time in trees; life was good, and food plentiful, so they had little reason to move. Something triggered their first radiation out of this comfort zone. Changing climate and vegetation almost certainly had a hand, and this was closely related to progressive closure of Tethys and consequent approach of Africa and Arabia to Eurasia. Not long after these continental connections were established, apes began their early migrations. By around 15–10 Mya they reached southwest Europe, central Europe and Asia – they had seen and skirted what remained of Tethys. But these were not our direct descendants.

The forerunners of the great apes are believed to have remained in their East African heartland until the time of their divergence into gorillas, chimpanzees and humans, although fossil evidence for this is extremely sparse. Biochemical studies of the DNA structure of living species have been used to determine the length of time needed to achieve the divergence observed. Chimpanzees and humans share around 98% of their DNA, and slightly less with gorillas. The molecular clock technique yields a three-way split between 7 and 5 million years ago to account for the 'small' differences – perhaps just before the time Tethys finally vanished.

I find this thought intriguing. It suggests that our closest ancestors, the ancient hominins could well have escaped from East Africa and reached the shores of Tethys just before the end, around 5.3 million years ago. Current wisdom based on the scant evidence that exists tends to confine these early hominins (also called australopithecines) to Kenya, Tanzania and Ethiopia. A short time later (3.5 to 2.3 Mya) they had reached as far as Lake Chad in the north-west and to Cape Province in the south, as demonstrated from their remains dating to this time.

However, I suggest that a combination of evolutionary and environmental changes prompted at least some of our distant relatives to leave the impoverished forests of East Africa some time before this, picking their way along the mighty River Nile and so reaching the Tethys in time to play along the shoreline before it recoiled into a hot inferno. The exclusive find that would confirm my proposition is yet to be made – perhaps it is now submerged just below sea level off the Nile Delta.

LAST GASP FOR TETHYS

There then followed catastrophe for the region and its life, an event so dramatic that its repercussions were felt throughout the world's oceans. What was it and how do we know? One of the best places to look for evidence of this event is about 40 minutes' drive north from Carboneras in the *Rio de Aguas* (River of Waters – and yes, it's unusual

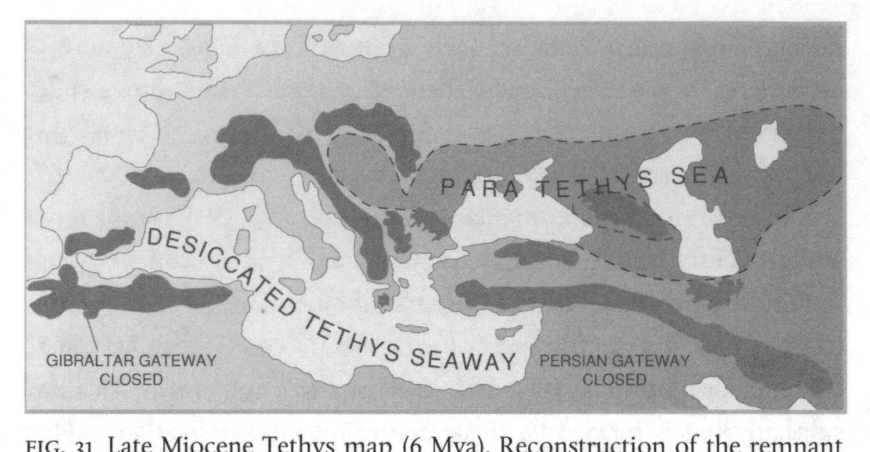

FIG. 31 Late Miocene Tethys map (6 Mya). Reconstruction of the remnant Tethys and Paratethys basins after closure and desiccation of the Tethys Seaway. Principal mountain ranges also shown.

to actually have water in the rivers in this region). This *mostly* dry river casts an almost complete meander loop around the ancient and attractive market town of Sorbas, such that the outer rim of houses is balanced precariously above 40-metre-high sheer cliffs of sandstone. Just below the sandstone, a couple of meander loops downstream, there are solid white cliffs a further 40 metres thick. These are made of gypsum, just some of the large expanse of gypsum across the region: cliff faces with myriad crystals sparkling brilliantly in the sunshine; individual transparent crystals that have grown to as much as 50 centimetres in length; twinned swallow-tail crystals, or rhomboids with the intriguing property of double-refraction – this means that when you peer through the rhomboidal crystal at, for example, a spot marked on a piece of paper, you actually see two spots side by side.

Gypsum is a very soft and soluble mineral. It can be scratched easily with your finger nail and readily carved into delightful ornaments. It can also be dissolved away by percolating rainwater to create large underground caverns and tunnels in exactly the same way as found below the surface in areas of limestone. Such well-developed gypsum karst scenery, as it is known, is rare to find anywhere in the world, which is why it has been thoughtfully preserved by the Spanish Government as a National Park in this part of the Sierra Alhamilla.

The gypsum is significant in another respect too, for it yields part of the answer to the question posed above. It is found not only in Spain, but right across the Mediterranean region today, all dating from exactly the same narrow band of time from 6.5 to 5.3 Mya – right at the end of the Miocene period. Gypsum is a mineral of calcium sulphate that is formed by the evaporation of seawater beyond its capacity to retain this chemical compound in solution. It is formed today in semi-isolated seas in arid parts of the world, such as the Aral Sea in Asia or along the margins of the Persian Gulf. Sodium chloride (common salt) is even more abundant in seawater, but is much more soluble than calcium sulphate, so that evaporation and concentration of chemical salts in the seawater must be even more pronounced before rock salt (halite) is precipitated out of solution. When the Deep Sea Drilling Project first penetrated deep beneath the Mediterranean Sea in the 1970s, a remarkable discovery was made. Thick deposits of both gypsum *and* halite were recovered in the cores that were brought to surface, and they too were late Miocene in age.

This was the final piece of evidence required. Ken Hsu, formerly Professor of Geology at Ohio State University, was chief scientist aboard that DSDP Leg to the Mediterranean. Ken is normally a lively man as well as an imaginative scientist, so I can well imagine his excitement at this new discovery. He announced, with suitable fanfare to the rest of the geological community and to the world at large, that the Mediterranean Sea had once been cut off from the world's oceans and so evaporated to dryness, leaving behind a dry carpet of salt deposits. When the Straits of Gibraltar opened, it was like opening the flood gates: the Mediterranean was refilled by a giant waterfall from the Atlantic Ocean. In the decades of research that followed, scientists found that this explanation is essentially correct albeit with some modifications.

The body of water that had evaporated was, in fact, the final remnants of the Tethys Ocean (Fig. 31). The end came in stages – slow and painful to the last. To the east, the Indian subcontinent had already collided with Asia throwing up the Himalaya. But still the Tethys remained open into the Early Miocene (20 Mya) with a long arm extending to the Persian Gulf and into what had become the Indian Ocean. A northern arm of Tethys spread into Asia through what are now the Black Sea, Caspian Sea and Aral Sea. Then some time between 16 and 12 million years ago, Arabia pushed further north and became firmly sutured together with central Asia along the line of the Zagros Mountains, creating the Middle East as we know it today. This finally closed the Tethys–Persian Gulf connection. As Africa continued to push northwards, the Taurus, Hellenides and Dinarides mountain ranges isolated its northern arm as a great inland body of water, called the Paratethys Sea.

By this stage Northwest Africa had very nearly closed in on Europe across the Straits of Gibraltar. All that was left of the once mighty Tethys was an irregularly shaped body of water roughly similar to the present-day Mediterranean Sea. Paratethys remained as a brackish-to-freshwater inland sea further to the north. The real trouble started with continued collision in the west, coupled with a marked drop in sea level as global temperatures cooled and polar ice built up. This finally isolated the Tethys completely from the Atlantic Ocean about 6.5 million years ago.

The climate at that time was even drier and more arid in those middle latitudes than it is today so that the influx of river water was unable to keep pace with the rate of evaporation. Progressively the level of the water fell, the Tethys subdivided into even more small sub-basins, and the rivers that still fed into these remnant basins incised deeply into their banks, cutting valleys across the former continental slope. Seismic reflection studies, carried out by oil companies currently prospecting in the eastern Mediterranean, show numerous examples of these valleys buried beneath the outer delta and slope off the River Nile.

As the Tethys waters evaporated, the concentration of sea salts increased many times over until normal marine life could no longer exist. Salt-tolerant species of red algae survived for longer, but mass death inevitably followed, and still there was no let-up in the burning heat. Natural salts began to precipitate from the concentrated brine pools, gypsum being at first most abundant and then, as the seas dried up completely, common rock salt (halite) lined a gaping, blistering white basin. Life as we know it, indeed life of any kind, was unable to exist where once there had been such profusion and diversity.

Complete evaporation of a sea the size that Tethys had become would leave a salt deposit about 25–30 metres thick over the deeper parts of the basin floors, and most of this would be halite rather than gypsum. This is easily calculated from the salinity of seawater and the volume of water contained in the basin. But, quite astoundingly, there are many hundreds of metres of *gypsum* preserved in Spain, Italy, Sicily, Crete, Cyprus and North Africa, in some cases up to 2000 metres thick. Add to this the thick sequences of both gypsum and halite drilled from beneath the sea-floor, and the numbers just don't add up. The Tethys before final closure at the Straits of Gibraltar would have been a similar depth to that of the Mediterranean now, and certainly no more salty. The only possible explanation for creating such a saline giant (as these deposits are now known) is that the last remnants of Tethys Ocean evaporated to dryness or near dryness many times over, being filled from the Atlantic by a tremendous cascade across the Strait of Gibraltar between each episode. This whole series of events lasted for around one million years, during which time all traces of the former Tethys waters were removed to the atmosphere and fell as rain on a different land draining to a new ocean. All the Tethyan salts were locked away as part of this saline giant, the origin of which had puzzled geologists for years.

Eventually a more permanent connection was re-established through the Straits of Gibraltar, but it was now the Mediterranean Sea that was filled to brimming. A new sea had been born where a former ocean had once thrived, and the new marine life that populated its waters was derived from the Atlantic Ocean. The turmoil of successive flooding and drying was finally over, a reign of peace and quiet ensued, and the Mediterranean slowly gained its present nature and azure calm. Barely a trace was left, save in the rocks, of this most amazing salinity crisis of all time. The Tethys Ocean had vanished forever.

11

Epilogue: Perspective on the Future

I know the highway along
which one age passed into another,
until fire or plant or liquid
was transformed into a deep rose,
into a spring of dense droplets,
into the inheritance of fossils.

From *Stones of the Sky*
by Pablo Neruda
(translated by James Nolan)

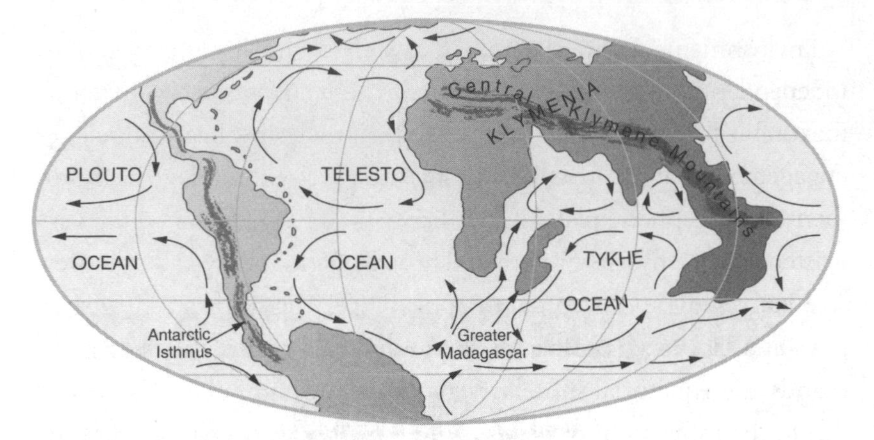

Projected global reconstruction at a time 50 million years into the future. New mountain ranges and ocean circulation is shown. The island chains to the west of Telesto Ocean and across Tykhe Ocean are island arcs formed behind new oceanic trenches and subduction zones.

For 250 million years of Earth history, the Tethys Ocean played a principal role in our colourful and evolving planet. The ocean varied much in its outline and extent but remained always at the heart of global events. Its influence was profound. Tethys nurtured revival and expansion of life after the great Permian extinction had wiped out 90–96% of all living species. The world quite simply began anew. Thereafter, Tethys was never far from the pulse of evolution – ichthyosaurs and ammonites, rudist reefs and nummulite shell banks. Species came and went, leaving tantalizing fossil remains that evoked past and very different worlds. But, each time new forms of life emerged and flourished, they were a little closer to those of today. Even the plankton gardens began to assume a more modern hue, feeding an ocean world that was alive and ever more familiar – with oysters and barnacles, crabs and lobsters, coral reefs and such a dazzling variety of fishes. Their evolution, life dramas and ultimate survival or extinction are all intricately bound to environmental factors in the ocean.

Environments within the Tethys, as well as globally, changed quite independently of the life they affected. These changes were driven by inexorable forces of nature, which had their own slow beat and rules of engagement that we are still working to unravel. Ultimately, the movement of lithospheric plates about Earth's surface determined the very existence of Tethys, its growth and broad extent, followed by progressive narrowing and eventual disappearance. The disposition of shallows and lagoons, continental slopes and mid-ocean ridges, tropical islands and equatorial shorelines, or temperate and polar seas, was all due to the movement of plates. With a mainly warm climate and the absence of polar ice caps, even the rise and fall of sea level was governed by the length and spreading rate of the mid-ocean ridge system.

Plate movement conspired to keep Tethys as a primarily low-latitude ocean throughout its existence. The rifting of Pangaea and growth of a new western arm of Tethys clean across the former supercontinent provided an equatorial passage through which ocean currents – both shallow and deep – circumnavigated the world. This pattern of ocean circulation helped to keep Earth as a greenhouse world with warmer climates and higher sea levels than those of today. It was not until the closure of Tethys was almost complete and the new north–south oriented Atlantic Ocean sufficiently broad that the global ocean was swept by an altogether different current system. Warm surface waters were carried in unprecedented volumes from equator to pole, where they cooled dramatically and then returned equator-wards in deep powerful bottom currents. This new conveyor-belt circulation, coupled with oceanic isolation of Antarctica over the South Pole, led directly to the most recent icehouse climate regime and hence to an alternation of glacial and interglacial conditions. The final stages of Tethys were witness to this latest plunge towards an ice age, and its demise was fully instrumental in what was to follow.

OCEAN REGULATOR

The ocean–climate nexus is fundamental today – just as it always has been. It controls the daily drama of Earth's weather systems and their distribution across the world. Overheating of the oceans (above 27°C) is directly responsible for the largest, most violent storms on Earth – hurricanes or tropical cyclones – and for their devastating effects on land and at sea. The energy released by a single tropical cyclone in one day would be enough to power the entire industrial production of the United States for one year. Winds drive the currents that redistribute heat from the equator to the poles, just as they are engaged themselves in the atmospheric heat engine. The transfer of heat energy in the ocean currents, every second of every day, is unbelievably staggering in its enormity. The Gulf Stream alone transfers some 2300 trillion joules per second of heat energy from the Caribbean towards NW Europe. The oceans operate as a giant thermostat, which regulates global temperature and, working in consort with the atmosphere, drives our weather systems.

Not only do the oceans hold heat energy, but they also provide one of the principal storehouses for carbon and carbon dioxide, and so they act to regulate greenhouse gases in the atmosphere. They are like a gigantic sponge, holding 50 times more carbon dioxide than the atmosphere, and they are thought to absorb 30–40% of the carbon dioxide produced by human activity. But the sea surface acts as a two-way control valve for gas transfer – opening and closing in response to gas concentration and ocean stirring. Carbon dioxide can be released back to the atmosphere and also removed more permanently into the sediment on the sea-floor. It is in this way that the oceans are responsible, at least in part, for longer-term changes in planetary climate.

At a recent meeting of the International Ocean Drilling Program in Bremen, Germany (2009), we were discussing just this strong ocean–climate link in the context of rapid climate change over the past few tens of millions of years. Professor Jim Zachos of the University of California in the USA outlined new and startling results from deep-sea drilling in the Arctic Ocean. He and his team had tracked the temperature of seawater in the past by studying changes in the proportion of different oxygen isotopes that were locked into the calcium carbonate of fossil plankton shells. This showed clearly that sea-surface temperature at the North Pole rose briefly and rapidly to 23°C in the middle Palaeogene Period around 55 million years ago – a quite dramatic difference from temperatures today, which hover at or below freezing point (0°C). This was during the last major flourish of the Tethys Ocean, when global sea level was estimated to be 150–200 metres higher than at present and Tethys waters flooded across much of Europe, central Asia and northern Africa. A great swathe of the sea swept across the Gulf of Mexico coastline and flooded into the western interior of America. From that time onwards, Tethys receded, patterns of ocean circulation changed, and global temperatures, on land and at sea, plummeted towards icehouse conditions.

We do not yet understand the trigger for that mid-Palaeogene rapid climate change, just what it was that put so much carbon dioxide into the atmosphere that the ocean regulator was forced to adjust its thermostat setting. My own bet is that it was closely related to the Greenland–Iceland–Faeroes superplume and excessive volcanic activity, which coincided more or less exactly in time. However, what we do know is that the *present* human-induced global warming has set the world once again on an upward thermal trend. The polar icecaps are melting, sea-surface temperatures are on the increase and sea level is rising. On a geological time frame this appears to be the start of a rapid change, although there is still huge debate about the short-term rate of increase. Jim Zachos' model suggests a return to temperatures similar to those of 33 million years ago by the end of this century, and then towards those of the mid-Palaeogene high by the year 2300. This dramatic scenario, of course, assumes no change in current human behaviour. However, for the present-day trend we *do* know the trigger for global warming – our profligate burning of fossil fuels. We have the potential to modify our behaviour and so reset the thermostat.

FUTURE OCEANS

Battle as we may with climate change, there is absolutely nothing that humans can do about the movement of plates across Earth's surface and hence about the profound environmental changes, good or bad, that will follow from their redistribution. We still do not even know how to predict the occurrence of volcanic eruptions or violent earthquakes associated with plate movement – although I am certain this knowledge will come within a few decades. But the thought of simply turning off the volcanoes and stopping the inexorable sliding of Earth's lithospheric plates into the oceanic trenches at subduction zones is a flight of fancy too far.

Nevertheless, it is possible and quite intriguing to forward track the known rates and directions of plate movement from today to, say, 50 million years into the future. This is exactly what I have done in order to construct the map at the beginning of this chapter. The result of this forward construction is by no means set in tablets of stone, as there is still far too much we do not know about the way that plate tectonics works – when and where a new mid-ocean ridge will form, for example, or when and where an old ocean will start to subduct back into the mantle. But, accepting that there are a number of such uncertainties, the world map I have drawn is a very reasonable picture of what the world may be like a great many years hence.

The northward movement of the southern continents, formerly elements of Gondwana, will undoubtedly continue unabated. Africa will force its way further into Europe and Arabia into central Asia so that the Mediterranean and all her component seas, together with the Black Sea, Caspian Sea and Aral Sea, will be completely squeezed out of existence. In their place, the current dispersed mountain chains of the region will rise higher and join up into a single range to rival the Himalaya. But the Himalaya themselves will not grow very much higher – and not simply because of increased rates of erosion. There are already signs that the force of India's impact into Asia is being taken up elsewhere. When we took the *Joides Resolution* drillship to the Indian Ocean in 1988, we saw clear evidence in seismic images beneath the sea-floor that the ocean crust was beginning to buckle and fold under the strain. Jim Cochran, a seismic expert from Lamont Doherty Geological Observatory and my co-chief scientist on that expedition, ventured to suggest that this might be the future site of a new submarine trench and subduction zone. Agreeing with his inference, I have therefore added one to my map – extending it roughly between the Maldives and Australia.

You will notice also from the map that Australia has moved north and fused with the islands of the western Pacific and South East Asia. This collision, coupled with slippage of the Indian plate into the Sumatran trench, will make the whole region into another new mountain chain – an extension from the Shan Shear Zone and the Himalayas in China. This will complete a gigantic mountain range stretching from coast to coast of the new supercontinent, matching the Central Pangaean Mountains that extended 7000 kilometres across the Pangaea some 250 million years before. I have called the supercontinent *Klymenia*, across which march the *Central Klymene Mountains*. It is no accident that these mark the final suture zone of the central and eastern Tethys Ocean. More fragments of the Tethys and her daughter seas – from the Mediterranean to the Aral Sea – will be thrust high into these peaks. More clues for geologists of that future time.

Doubtless the new Klymenia, too, will eventually begin to rift and break apart – but that is still further into the future. In the meantime, there is already a major rift system today that will almost certainly begin to open into a full ocean with a mid-ocean ridge spreading centre. This is the East African, Red Sea and Gulf of Aden rift system, which radiates out from the Ethiopian hotspot that I mentioned very briefly at the end of Chapter 2. The East African rift marks the very beginning of continental break-up. The crust beneath the rift is hot with much volcanic activity, relatively thinner than normal continental crust and already stretching apart. Eventually it will founder and the sea will flood in, which is exactly what happened in the case of the Red Sea. This is truly a very narrow incipient ocean in the making. Its deep axial region is being forced further apart as new ocean crust wells up. There are submarine hot springs and metal-rich sediments – perhaps even black smoker chimneys that we have not yet discovered. The Gulf of Aden already has a spreading centre, which is the continuation of the Indian Mid-Ocean Ridge. The whole region is extremely exciting and geologically unique – an area from which we can learn much about the splitting of continents and formation of oceans. Indeed, it is a perfect analogue for early Jurassic time when the western tip of the Tethys Ocean first cut a rift through central Pangaea.

On my map of the future world, therefore, I have shown Africa split into two parts along the former East African Rift such that the region east of the rift has drifted off into the former Indian Ocean and fused with Madagascar. The ocean in between – the new *Tykhe Ocean* – is shown as already over 1000 kilometres wide. In fact, it could even be double this width if spreading started almost immediately (say, a million years from now) and rates were around 4 centimetres per year, which is only slightly faster than Atlantic spreading today. The former Red Sea I have shown opening in a wedge shape, as the Central Klymene Mountains would have temporarily halted its incision further north.

To the west of Klymenia the former Atlantic Ocean will continue to widen until it comes to dominate the planet – the *Telesto Ocean* I have named it. As the ocean widens still further, the old, cold ocean crust furthest from the spreading ridge will become so dense and heavy that it breaks through and begins to sink back into the mantle – a new subduction zone will have formed. We really do not know when or where this might occur, so I have suggested that a string of trenches and associated volcanic island arcs have developed along the Americas on the western margin of Telesto.

Note also that Antarctica will have drifted north by this stage and fused with South America. There will be a single elongate continent – *Metisa* – that stretches from pole to pole, as part of the former Antarctica remains over the South Pole, while part of Greenland drifts across the North Pole. There will be permanent ice caps at both ends of Metisa and a long mountainous backbone along its western shores abutting the *Plouto Ocean*. Amongst all geological reconstructions of past continents and oceans, there has never been one quite like this. Perhaps this is not surprising, for although cycles of change are forever a part of Earth's long history, the past is never exactly repeated. The elongate Metisa will provide an effective barrier to latitudinal circulation. Instead, there will be a strong meridional current system, with powerful cold-water bottom currents generated in the polar regions sweeping towards the equator, and expansive regions of intense upwelling creating a very fertile ocean. Climatic zonation will be pronounced, from very cold to very hot. The Central Klymene Mountains will create a more extensive region, marked by monsoonal climate, than that which exists today. Given such a wide variety of ecological niches, life will flourish as never before, though humans may have long disappeared.

When Alfred Lord Tennyson wrote his prophetic lines in *Locksley Hall* 'For I dipt into the future far as human eye could see, / Saw the vision of the world and all the wonders there would be', he can scarcely have envisioned anything as 'unlikely' or as far distant as the world I have described above. However, I hasten to assert that this future world is by no means an imaginary one. Africa *is* moving north and the Mediterranean *will* be replaced by mountains. The East African Rift and Red Sea are *already* rifting and spreading to form a new ocean. The Atlantic *is* steadily growing wider, and so on. Ocean circulation is determined by the disposition of continents and oceans and is a key driver for global climate. There are, of course, other factors that influence climate and life on Earth, such as changing astronomical orbits, and superplume volcanic eruptions. These too will play their part in Earth's future and enrich the detail of my world map 50 million years from now.

The names I have chosen for the new oceans and continents of that world are selected with due care from Greek mythology. They are the names of the *Oceanides,* the children of Tethys and Okeanos, and so make a fitting tribute to a former great ocean that reigned supreme for 250 million years of Earth history. The Tethys Ocean not only shaped the splendour and dynamism of the world we know today but from its study we gain a new and significant perspective of our own place in the world and in the grand vista of time. I believe that this perspective can also help engender an appropriate sense of responsibility, pride and humility as humankind faces the enormous challenges of the present and future.

SUGGESTED FURTHER READING

Beerling, D., 2007, *The Emerald Planet*, Oxford University Press

Benton, M. J., 2003, *When Life Nearly Died: The Greatest Mass Extinction of All Time*, Thames and Hudson

Bjornnerud, M., 2005, *Reading the Rocks: The Anatomy of the Earth*, Westview Press

Byatt, A., Fothergill, A., and Holmes, M., 2002, *The Blue Planet: A Natural History of the Oceans*, BBC/DK

Conway Morris, S., 2004, *Burgess Shale*, Oxford University Press

Dawkins, R., 1996, *Climbing Mount Improbable*, Viking Press

Dixon, D., Jenkins, I., Moody, R., and Zhuravlev, A., 2001, *Cassell's Atlas of Evolution*, Cassell & Co.

Fortey, R., 1999, *Life: A Natural History of the First Four Billion Years of Life on Earth*, Vintage Books

Fortey, R., 2001, *Trilobite: Eyewitness Guide to Evolution*, Harper Collins

Fortey, R., 2004, *Earth: An Intimate History*, Knopf Publishing/Random House

Holland, H. D., and Petersen, U., 1995, *Living Dangerously: The Earth, its Resources and the Environment*, Prentice Hall

Jones, S., 2001, *Almost Like a Whale*, Black Swan

Kunzig, R., 2000, *Mapping the Deep: The Extraordinary Story of Ocean Science*, Sort of Books

Marshak, S., 2005, *Earth: Portrait of a Planet* (2nd edition), W. W. Norton & Co.

Monroe, J. S., and Wicander, R., 2001, *The Changing Earth: Exploring Geology and Evolution*, Brooks/Cole

Nield, T., 2007, *Supercontinent: Ten Billion Years in the Life of Our Planet*, Granta Books

Pickering, K. T., and Owen, L. A., 1997, *An Introduction to Global Environmental Issues* (2nd edition), Routledge

Pinet, P. R., 1996, *Invitation to Oceanography*, West Publishing Co.

Press, F., and Siever, R., 2001, *Understanding Earth* (3rd edition), W. H. Freeman.

Redfern, R., 2000, *Origins: The Evolution of Continents, Oceans and Life*, Cassell & Co.

Southwood, R., 2003, *The Story of Life*, Oxford University Press

Stanley, S. M., 1989, *Earth and Life Through Time* (2nd edition), W. H. Freeman

Stewart, I., 2005, *Journeys from the Centre of the Earth*, Century

Stow, D. A. V., 2004, *An Encyclopedia of the Oceans*, Oxford University Press

Stow, D. A. V., 2005, *Oceans: An Illustrated Reference*, University of Chicago Press

Thurman, H. V., and Trujillo, A. P., 1999, *Essentials of Oceanography* (6th edition), Prentice Hall

Tudge, C., 2000, *The Variety of Life: A Survey and a Celebration of All the Creatures That Have Ever Lived*, Oxford University Press

Van Andel, T. H., *New Views on an Old Planet: A History of Global Change*, Cambridge University Press

Walker, G., and King, D., 2008, *The Hot Topic: How to Tackle Global Warming and Still Keep the Lights On*, Bloomsbury

Wilson, E. O., 1992, *The Diversity of Life*, Penguin

Zalasiewicz, J., 2008, *The Earth After Us: What Legacy Will Humans Leave in the Rocks?* Oxford University Press

GLOSSARY OF TERMS

ADAPTIVE RADIATION The effect of natural selection whereby the genes best suited for survival in a gene pool become more common within a species population.

ANAEROBIC Lacking oxygen. Also refers to organisms that do not depend on oxygen for respiration.

ANOXIC The absence of free oxygen (O_2).

ARTHROPOD An animal belonging to the phylum arthropoda; characterized by jointed appendages and a hard exterior covering (exoskeleton); includes all insects and crustaceans.

ASTHENOSPHERE The hot, soft region of upper mantle that lies directly below the lithosphere.

BASALT A dark, fine-grained, volcanic igneous rock composed of minerals enriched in ferromagnesian silicates; it typifies the oceanic crust.

BATHYMETRY The measurement of depth below sea level in the ocean in order to delineate the submarine topography.

BENTHIC Pertaining to the ocean bottom.

BLACK SHALE Organic-carbon-rich sedimentary rock, typically formed in an anoxic environment.

BLACK SMOKER Deep-sea hydrothermal vent (hot spring), mainly occurring at mid-ocean ridges and in back-arc basins.

BOTTOM WATER A general term applied to dense water masses that sink to the floor of ocean basins.

BRECCIA Coarse-grained sedimentary rock composed mainly of angular rock fragments, finer-grained matrix and a mineral cement (see also 'conglomerate').

CALCAREOUS Composed of calcium carbonate ($CaCO_3$).

CARBONATE COMPENSATION DEPTH (CCD) The depth in the ocean below which material composed of calcium carbonate is dissolved and therefore does not accumulate on the sea-floor.

CENOZOIC Of, belonging to or designating the latest era of geological time, from 65 million years ago to the present.

CEPHALOPOD An animal belonging to the molluscan class cephalopoda, which includes squids, octopods, cuttlefish and nautiloids.

CHORDATE An animal belonging to the phylum chordata, characterized by a notochord, a dorsal, hollow nerve chord, and gill slits, all of which appear at least some time during the animal's lifecycle.

CLAST An individual fragment of rock, mineral or fossil included as part of a sedimentary rock.

COCCOLITH A microscopic calcitic skeletal platelet that helps protect certain marine phytoplankton (coccolithophores or nannofossils); the dominant component of certain limestones, such as chalk.

CONGLOMERATE Coarse-grained sedimentary rock composed mainly of rounded rock fragments, finer-grained matrix and a mineral cement (see also 'breccia').

CONTINENTAL MARGIN The region between the part of a continent above sea level and the deep ocean floor. This is commonly subdivided into the continental shelf, the slope and the rise.

CORAL REEF A marine ridge or mound composed predominantly of compacted coral, together with algal material and biochemically deposited magnesium and calcium carbonates.

CORE The innermost region of the Earth, beginning at a depth of around 2900 km; thought to be mainly composed of iron and nickel; divided into outer liquid core and inner solid core.

CORIOLIS FORCE The natural effect that induces the deflection of the path of winds and ocean currents (or other free-moving bodies) as a result of the rotation of the Earth.

CRATON The relatively very old and stable part of a continent; its original nucleus.

CROSS BEDDING Inclined layers or laminae in a sedimentary rock that were formed by currents of wind or water during deposition.

CRUST The outermost, thinnest and coolest layer of the Earth. Consists mainly of either granite (continental crust) or basalt (oceanic crust). Thickness varies from 5 to 70 km.

CRYPTOZOIC Early period of time from the first beginnings of life on Earth to the start of the Cambrian period; means 'hidden life', as fossils are rarely preserved. More or less synonymous with Pre-Cambrian.

DEBRIS FLOW The rapid down-slope flowage of unconsolidated material, typically comprising an unsorted mix of large boulders, small pebbles and sand supported in a mud matrix.

DENSITY The mass per unit volume of a substance.

DENUDATION The exposure and removal of rock by the action of flowing water.

DETRITUS Either *organic* matter, such as animal wastes and bits of decaying tissue, or *inorganic* sedimentary material (= sediment).

DIATOM Photosynthetic, unicellular protists that belong to the phylum chrysophyta; they possess a glassy covering composed of silica.

DIKE A tabular or sheet-like intrusion of igneous rock that cuts across bedding and other structures in the surrounding rocks.

DINOFLAGELLATE Photosynthetic, usually single-celled protists, possessing two flagella, belonging to the phylum pyrrophyta.

ECHINODERM An animal belonging to the phylum echinodermata, including sea stars, brittle stars, sea urchins, sea cucumbers and crinoids.

EVAPORITE A type of sediment precipitated from a concentrated aqueous solution, usually by the evaporation of water from a basin with restricted circulation; includes halite, gypsum and anhydrite.

FAMILY A type of taxon, more exact than orders, and comprising several or many different genera.

FAULT Natural break or rupture between rocks along which movement takes place; such movement is one of the principal causes of earthquakes.

FOOD CHAIN A sequence of feeding relationships among a group of organisms that begins with producers and continues in a linear fashion to higher-level consumers; generally a simplified picture of reality.

FOOD WEB A representation of the complex feeding networks, comprising many food chains or part chains, which exist in an ecosystem.

FORAMINIFERAN Amoeba-like protozoan (single-celled organism), many of which produce an elaborate shell of calcium carbonate; also known as foram and foraminifer.

FRACTURE ZONE A linear zone of highly irregular, faulted topography that is generally oriented perpendicular to ocean spreading ridges; also known as a transform fault; the term can be used for a region of major faulting on land.

GABBRO A dark, speckled, coarse-grained igneous rock formed at depth within the crust, composed chiefly of plagioclase feldspar and pyroxene minerals; the intrusive equivalent of basalt.

GASTROPOD A member of the molluscan class gastropoda, including snails, limpets and abalones.

GONDWANA One of the principal former continents, composed of the present day continents of South America, Africa, Antarctica, Australia and India, as well as parts of other continents including southern Europe, Arabia and Florida.

GRANITE A light-coloured, coarse-grained, intrusive igneous rock formed deep within the crust, composed mainly of quartz and feldspar minerals; the dominant rock type of continental crust.

GREENHOUSE EFFECT The warming of the Earth's atmosphere owing to the build-up of greenhouse gases, such as carbon dioxide and methane.

GULF STREAM Warm ocean current that originates in and around the Caribbean and flows across the North Atlantic to north-west Europe.

GYRE A large water circulation system of geostrophic currents rotating clockwise (northern hemisphere) or anticlockwise (southern hemisphere).

HOTSPOT Localized zone of melting and upwelling in the asthenosphere/lithosphere above which volcanic activity is abundant.

HYDROCARBON Organic compounds composed of hydrogen, carbon and oxygen; the main components of petroleum (both oil and gas).

HYDROTHERMAL VENT An opening from which emanates hot, metal-enriched water; commonly found at mid-ocean ridge sites.

ICEHOUSE EFFECT The cooling of the Earth's average global temperature occurring during an ice age, commonly due to an increase in albedo.

IGNEOUS ROCK Any rock formed by cooling and crystallization of magma (molten rock material); one of the three main groups of rock.

INTERGLACIAL PERIOD Period of time during an ice age when temperatures increase leading to glacier retreat.

ISLAND ARC A chain of volcanic islands associated with oceanic subduction zones, lying on the continent side of deep-sea trenches. Formed by the partial melting of the lithosphere as a plate is subducted.

ISOTOPE Different form of the same element related to variations in the number of neutrons in the nucleus.

KRILL Pelagic, shrimp-like creatures that belong to the arthropod order euphausiacia.

LAURASIA A former northern hemisphere continent composed of the present-day continents of North America, Greenland, Europe and Asia.

LAVA Magma extruded at the Earth's surface; used for both the still-molten form and when hardened into volcanic rock.

LAW OF SUPERPOSITION States that strata (sediment layers) lower in a sequence must be older than those higher up, provided that major earth movements have not overturned the succession.

LITHIFICATION The process by which loose, unconsolidated sediment is compacted and cemented into a sedimentary rock.

LITHOSPHERE The relatively cool, brittle outer shell of the Earth, including the crust and upper mantle. Extends to a depth of around 100 km.

LITTORAL ZONE A subdivision of the benthic province between the high- and low-tide marks, equivalent to the intertidal zone.

MAGMA CHAMBER A reservoir of molten rock within Earth's otherwise solid lithosphere.

MAGNETIC REVERSAL A complete reversal of the north and south magnetic poles. This has occurred at irregular intervals throughout geological time.

MANGROVE A dense growth of salt-tolerant mangrove trees and shrubs in marsh-like shoreline environments of the tropics and subtropics.

MANTLE The thick layer of the Earth's interior between the crust and the core, composed of ferromagnesian silicate minerals.

MANTLE PLUME A stationary column of magma originating deep within the mantle and rising to the Earth's surface to form volcanoes or basalt plateaus; it underlies hotspots; it is called a 'superplume' where it is particularly long-lived and has erupted very large volumes of igneous rock.

MASS EXTINCTION A catastrophic, widespread perturbation where major groups of species become extinct in a relatively short period of geological time.

MELANGE A mixture, especially of rock types.

MESOZOIC Of, belonging to or designating the era of time including the Triassic, Jurassic and Cretaceous periods, between 245 and 65 million years ago.

METAMORPHIC ROCK Any rock altered by high temperature or pressure and the chemical activities of fluids; one of the three main classes of rocks.

MID-OCEAN RIDGE A long mountain range that forms along cracks in the ocean floor where magma breaks through the Earth's crust; the site of formation of new ocean crust and sea-floor spreading as two tectonic plates move apart; also known as a divergent plate margin.

MILANKOVITCH CYCLE Cyclic variations in climate, with regular periodicities of around 20,000, 40,000 and 100,000 years, as a result of irregularities in the Earth's rotation and orbit.

MINERAL A naturally occurring, inorganic, crystalline solid having characteristic physical properties and a narrowly defined chemical composition.

MOHOROVIÇIC DISCONTINUITY A compositional and density discontinuity marking the interface between the rocks of the crust and the mantle; also known as the 'Moho'.

MONSOON A regional-scale wind system that predictably changes direction with the passing of the seasons; mostly in south and south-east Asia; summer monsoons are often accompanied by heavy precipitation.

NICHE An organism's role in its environment.

OCEANIC CRUST The outermost shell of the Earth, around 5–10 km thick, that underlies oceans, composed of the basic igneous rocks – basalt, dolerite and gabbro – and commonly overlain by sedimentary layers.

OCEANIC GATEWAY Partial topographic barrier between ocean basins, through which a deeper, narrow zone or channel allows the interchange of water masses.

OIL TRAP A structural or stratigraphic feature within sedimentary rocks of the Earth's crust that confines oil to a certain area.

OPHIOLITE A sequence of igneous rocks, thought to represent a fragment of oceanic lithosphere now emplaced onto the continent, composed of peridotite overlain successively by gabbro, sheeted dolerite dikes and pillow basalts.

ORDER A type of taxon, more exact than a class, for example primates.

OROGENY The process of forming mountains, including folding, thrust faulting and uplift; an episode of mountain building.

PALAEOCEANOGRAPHY The study of past oceans, their location and characteristics.

PALAEOZOIC Of, belonging to or designating the era of time including the Cambrian, Ordovician, Silurian, Devonian, Carboniferous and Permian periods, between 545 and 245 million years ago.

PANGAEA (also PANGEA) The single supercontinent of the late Palaeozoic and early Mesozoic Eras that comprised all of the present-day continents.

PANTHALASSA The late Palaeozoic to early Mesozoic worldwide ocean that surrounded the supercontinent Pangaea.

PELAGIC Of the ocean; more commonly the open ocean away from the shoreline.

PHANEROZOIC Of, belonging to or designating the period of time (the eon) including the Palaeozoic, Mesozoic and Cenozoic Eras, extending from 545 million years ago to the present.

PHOTOSYNTHESIS The process by which some organisms use the energy of sunlight to produce organic molecules, usually from carbon dioxide and water.

PHYLUM A type of taxon, more exact than kingdom, and less than class.

PHYTOPLANKTON Tiny photosynthetic organisms that float near the ocean surface.

PILLOW LAVA Bulbous masses of basalt, somewhat resembling pillows, formed when basalt is rapidly chilled under water.

PLANKTON Animals and plants that float passively in the ocean.

PLANKTON BLOOM The sudden and rapid multiplication of plankton that results in dense concentrations of planktonic organisms.

PLATE TECTONICS The movement of large segments of the Earth's outer crust and mantle (lithospheric plates) relative to one another; the new paradigm of the earth sciences developed through the 1960s and 1970s.

PLUTON A large, irregularly shaped body of igneous rock, originally intruded within other rocks deep beneath Earth's surface.

POPULATION A group of individuals of the same species that occupies a specified area.

PRIMARY PRODUCER Those organisms in a food chain, such as green plants and photosynthesizing or chemosynthesizing bacteria, upon which all other members of the food chain depend directly or indirectly; those organisms not dependent on an external source of nutrients; also known as autotrophs.

PRIMARY PRODUCTIVITY The quantity of organic matter that is synthesized from inorganic materials by autotrophs.

RADIOACTIVE DECAY The spontaneous decay of an atom to an atom of a different element by emission of one or more particles or photons from its nucleus (alpha, beta and gamma decay).

RADIOLARIAN A protozoan that has an intricate shell made of silica and that uses pseudopods to capture prey.

RED BED A sedimentary sequence of reddish-coloured, iron-stained rocks deposited in continental environments.

REGRESSION A relative fall in sea level resulting in the deposition of terrestrial strata over marine strata.

RESERVOIR ROCK Sedimentary strata having porous and permeable properties and acting as a reservoir for hydrocarbons.

RIFT VALLEY The fault-bounded valley found along the crest of many ocean ridges, created by tensional stresses that accompany the process of sea-floor spreading; also occurs on continents.

SALINITY A measure of the total concentration of dissolved solids in seawater, generally expressed in parts per thousand.

SALT DOME A columnar intrusion of salt through a sedimentary succession, commonly deforming the strata upwards into an anticlinal fold.

SALT PAN A shallow basin in the ground, typical of the coastal zone in arid regions, where salt water is evaporated by the heat of the sun.

SEAGRASS A type of marine plant most similar in structure to land plants.

SEA ICE Frozen seawater.

SEA-FLOOR SPREADING The process by which oceanic crust is created at the crest of ocean ridges, and tectonic (lithospheric) plates diverge.

SEAMOUNT A submarine mound, usually of volcanic origin, that rises sharply from the sea-floor.

SEDIMENT LOAD The amount of sediment carried by air, water or ice.

SEDIMENTARY BASIN Thick accumulation (typically 1–15 km thick) of sediments in a particular region, occurring below both the continents and the sea-floor.

SEDIMENTARY CYCLE The process by which sediments are formed, transported, deposited and then compacted and cemented into sedimentary rocks, before being uplifted and once more subjected to weathering and erosion.

SEDIMENTARY ROCK A rock formed from the compaction and cementation of sediment; one of the three principal rock types.

SEISMIC Pertaining to a naturally occurring or artificial earthquake or earth vibration.

SEISMIC SURVEYING The use of sonar and explosive devices and measuring equipment to study the nature of natural earthquakes and artificial vibrations.

SEXUAL REPRODUCTION The process by which two parent organisms produce an offspring by the fusion of sex cells produced by each parent.

SHELF SEA The area of ocean found at the edge of the continental shelf, before the continental slope.

SILICATE A mineral containing derivatives of silica (SiO_2) as the principal component.

SILICEOUS Material whose composition is silica (SiO_2); generally referring to biogenically produced siliceous material (from diatoms, radiolarians, etc).

SLIDE A type of down-slope movement of material along one or more surfaces of failure; can be both subaerial and subaqueous in occurrence.

SLUMP A type of down-slope movement of material that takes place along a curved surface of failure and results in backward rotation of the slump mass; more contorted than a slide.

SPECIES A population of similar individuals that can reproduce and produce fertile offspring.

SPREADING CENTRE Mid-oceanic ridges where sea-floor spreading occurs.

STRATIGRAPHY The branch of geology that studies the age relationship and significance of layered sedimentary rocks and the sequence of fossils they contain; also known as historical geology.

SUBDUCTION The movement of one lithospheric plate underneath another so that the descending plate is consumed into the mantle; generally creates a deep oceanic trench at the point of descent.

SUBDUCTION ZONE An area where subduction is occurring.

SUBMARINE CANYON Deeply incised, steep-walled valley, commonly V-shaped in profile, that cuts into the rocks and sediments of the outer continental shelf and the continental slope.

SUBMARINE FAN A cone-shaped sedimentary deposit that accumulates on the continental slope and rise; generally fed from a distinct point or line-source of sediment such as a major river or delta.

SUBSIDENCE The sinking of large portions of the Earth's crust under the influence of major tectonic forces.

SUPERPLUME See 'mantle plume'.

SYMBIOSIS An intimate living relationship between two different organisms.

TALUS SLOPE A slope in a subaerial or submarine setting where weathered, eroded and fragmented material collects.

TECTONIC UPLIFT A rise in topographic height due to crustal movement caused by major tectonic forces.

TETHYS OCEAN An immense ocean that separated Gondwanaland from Laurasia – the subject of this book.

THERMOHALINE CIRCULATION The movement of water masses that results from differences in density caused by differences in salinity and/or temperature.

TRACE ELEMENTS Elements found in small quantities, typically less than 1 part per million.

TRANSFORM FAULT A steep boundary separating two lithospheric plates along which there is lateral slippage.

TRANSGRESSION The incursion of the sea over land areas, resulting in the deposition of marine over terrestrial sediments.

TRENCH Long, narrow and deep topographic depression associated with a volcanic arc, that together mark a collision or subduction zone where one lithospheric plate is overriding another.

TSUNAMI A destructive sea wave that is usually produced by an earthquake but can also be caused by submarine landslides or volcanic eruptions.

TURBIDITE Sediment layer, typically with graded bedding, that is deposited by a turbidity current.

TURBIDITY CURRENT A density-driven current of sediment-laden water that flows swiftly down-slope, in some cases travelling many hundreds of kilometres from the shelf edge to the abyssal plain.

UPWELLING The slow upward transport of water to the surface from depth; generally recycling nutrient elements and organic material and so leading to increased primary productivity in surface waters.

ZOOPLANKTON Animal plankton such as foraminifera and radiolaria.

ZOOXANTHELLAE Symbiotic dinoflagellates that live in the tissue of corals and other reef-building organisms.

INDEX

Index

Index